MATHEMATICAL MODELING

Industrial Mathematics is growing enormously in popularity around the world. This book deals with REAL industrial problems from REAL industries. Presented as a series of case studies by some of the world's most active and successful industrial mathematicians, this volume shows clearly how the process of mathematical collaboration with industry can not only work successfully for the industrial partner, but also lead to interesting and important mathematics.

The book begins with a brief introduction, where the equations that most of the studies are based upon are summarized. Thirteen different problems are then considered, ranging from cooking of cereal to the analysis of epidemic waves in animal populations. Throughout the work, the emphasis is on providing to people in industry information that they can use.

This book is suitable for all final year undergraduates, master's students, and Ph.D. students who are working on practical mathematical modeling.

Ellis Cumberbatch is Joseph H. Pengilly Professor of Mathematics and Director of the Mathematics Clinic at Claremont Graduate University. He has held positions as Lecturer in Mathematics at Leeds University and Professor of Mathematics at Purdue University and he was a Senior Visiting Fellow at Oxford University.

Alistair Fitt is a Professor in the Faculty of Mathematical Studies at the University of Southampton. He has attended more than forty industrial study groups and student meetings in Europe, the United States, Canada, Asia, Mexico, Brazil and China and has close links with many different industries.

MATHEMATICAL MODELING

Case Studies from Industry

Edited by

ELLIS CUMBERBATCH
Claremont Graduate University

ALISTAIR FITT
University of Southampton

CAMBRIDGE
UNIVERSITY PRESS

PUBLISHED BY THE PRESS SYNDICATE OF THE UNIVERSITY OF CAMBRIDGE
The Pitt Building, Trumpington Street, Cambridge, United Kingdom

CAMBRIDGE UNIVERSITY PRESS
The Edinburgh Building, Cambridge CB2 2RU, UK
40 West 20th Street, New York, NY 10011-4211, USA
10 Stamford Road, Oakleigh, VIC 3166, Australia
Ruiz de Alarcón 13, 28014 Madrid, Spain
Dock House, The Waterfront, Cape Town 8001, South Africa

http://www.cambridge.org

First published 2001

Printed in the United Kingdom at the University Press, Cambridge

Typeface Times Roman 10.25/13 pt.   *System* LATEX 2$_\varepsilon$ [KW]

A catalog record for this book is available from the British Library.

Library of Congress Cataloging in Publication Data

Mathematical modeling: case studies from industry/edited by Ellis Cumberbatch,
Alistair Fitt
    p. cm.
Includes bibliographical references and index.
ISBN 0 521 65007 0   ISBN 0 521-01173-6 (pb.)
1. Mathematical models. – Industrial applications.   I. Cumberbatch, Ellis.
II. Fitt, Alistair, 1957-

T57.62.M38 2001
670′.1′.5118 – dc21   00-054672

ISBN 0 521 65007 0 hardback
ISBN 0 521 01173 6 paperback

Contents

Contributors   *page* xi
Preface   xv
Introduction   1

0 Mathematical Preliminaries   **5**
  0.1 The Continuum Model   5
     0.1.1 Conservation Laws   5
  0.2 Diffusion   6
  0.3 Flow and Deformation of Solids and Fluids   9
     0.3.1 Equations of Motion   10
  0.4 Incompressible Linear Viscous Fluid   11
     0.4.1 Inviscid Incompressible Fluid   13
     0.4.2 Inviscid Incompressible Irrotational Flow   14
     0.4.3 Compressible Viscous Flow   14
     0.4.4 Compressible Inviscid Flow   15
     0.4.5 The Reynolds Number and Flow in a Viscous
           Boundary Layer   16
     0.4.6 Viscous Flow at Low Reynolds Number 17
     0.4.7 Viscous Flow in a Thin Layer   18
  0.5 Linear Elasticity   19
  0.6 Summary   20
  References   20

1 Fluid-Mechanical Modelling of the Scroll Compressor   **22**
  Preface   22
  1.1 Introduction   22
  1.2 Leakage between Chambers   25
     1.2.1 Governing Equations and Boundary Conditions   26
     1.2.2 Dimensionless Parameters   28
     1.2.3 Dimensionless Equations   30
     1.2.4 Solution in the Quasi-Steady Limit   32

1.3   Conservation Equations for the Chambers   33
1.4   The Coupled Problem   34
    1.4.1   The Small Coupling Limit   36
1.5   Numerical Results   37
1.6   Conclusions   43
Acknowledgements   44
References   44

2   **Determining the Viscosity of a Carbon Paste Used in Smelting**   **46**
Preface   46
2.1   Continuous Electrode Smelting   46
2.2   Problem Formulation   49
2.3   Simplified Analysis   51
    2.3.1   Corner Solutions   51
2.4   Special Geometries   54
    2.4.1   Further Analysis of the Velocity Test   56
    2.4.2   Analysis of the Viscometer Test   57
    2.4.3   Analysis of the Plasticity Test   58
    2.4.4   The Boundary Layer at the Base of the Sample   58
2.5   Numerical Analysis and Results   60
    2.5.1   Finite-Element Method   60
    2.5.2   Results   61
2.6   Final Conclusions   63
Acknowledgement   64
References   65

3   **The Vibrating Element Densitometer**   **66**
Preface   66
3.1   Introduction   66
3.2   Resonance   70
3.3   Added Mass Model   71
3.4   Fluid–Plate Model   72
    3.4.1   Plate Equation   72
    3.4.2   Fluid Equation   73
    3.4.3   Fluid–Plate Interaction   73
3.5   Simple Analysis: Incorrect Boundary Conditions   74
3.6   Solution with Clamped Boundary Conditions   76
3.7   Remarks   77
Appendix   78
References   79

4 Acoustic Emission from Damaged FRP-Hoop-Wrapped Cylinders **80**
Preface 80
4.1 Introduction 80
4.2 Problem Description 83
4.3 Problem Solution 89
4.4 Further Analysis 91
4.5 Conclusion 94
References 96

5 Modelling the Cooking of a Single Cereal Grain **97**
Preface 97
5.1 Introduction 97
5.2 The Problem 98
5.3 Background 99
5.4 Heating a Single Grain 100
 5.4.1 Sphere 102
 5.4.2 Ellipsoid 102
5.5 Timescales for Wetting and Heating – Linear Models 103
5.6 Wetting the Grains – a Nonlinear Model 104
 5.6.1 Numerical Solutions 105
 5.6.2 Analytic Solutions – Mean Action Time 106
 5.6.3 Log Mean Diffusivity 108
 5.6.4 Degree of Overcook for the Present Process 109
5.7 Temperature Dependence of Wetting Times 111
5.8 Sensitivity Analysis 112
5.9 Conclusions and Further Extensions 112
Acknowledgements 114
References 114

6 Epidemic Waves in Animal Populations: A Case Study **115**
Preface 115
6.1 History of RHD and its Introduction into New Zealand 116
6.2 What is Known about the Disease 117
6.3 What We Want to Know 117
6.4 The Modelling. Analytical/Numerical 118
 6.4.1 Case: No Immunity ($R(x, t) = 0$) and No Breeding ($a = 0$) 120
 6.4.2 Case: No Immunity ($R(x, t) = 0$) But Breeding Season ($a \neq 0$) 125
 6.4.3 Parameter Values 127

6.5	Immunity	130
6.6	Results and Conclusions	131
6.7	Further Work	132
	References	133

7 Dynamics of Automotive Catalytic Converters — **135**
7.1	Introduction	135
7.2	Model Equations	137
7.3	Single-Oxidand Model and Nondimensionalization	141
7.4	Asymptotic Analysis of the Single-Oxidand Model	144
	7.4.1 Warm-up Behavior	145
	7.4.2 Light-off Behavior	147
7.5	Numerical Methods and Results	150
7.6	Further Analysis of the Single-Oxidand Model	154
7.7	Concluding Remarks	157
	Acknowledgements	158
	References	158

8 Analysis of an Endothermic Reaction in a Packed Column — **160**
8.1	Introduction	160
8.2	The Problem and the Model	161
8.3	Analysis	166
8.4	Discussion	175
8.5	Further Modelling Considerations	176
	References	178

9 Simulation of the Temperature Behaviour of Hot Glass during Cooling — **181**
9.1	Cooling of Glass	182
9.2	Mathematical Formulation of the Problem	182
	9.2.1 Heat and Radiative Transfer Equations	182
	9.2.2 Modelling of the Boundary Conditions for the Heat Transfer Equation	184
9.3	Numerical Solution Methods	187
	9.3.1 The Heat Transfer Equation	187
	9.3.2 Ray Tracing	187
	9.3.3 A Diffusion Approximation	189
	9.3.4 Two-Scale Analysis	190
9.4	Numerical Simulation and Results	195
9.5	Conclusions and Further Questions	196
	References	197

10 Water Equilibration in Vapor-Diffusion Crystal Growth **199**
Preface 199
10.1 Introduction 199
10.2 Formulation 203
10.3 Analytical Treatment 209
    10.3.1 Geometry 209
    10.3.2 Method of Multiple Timescales 210
    10.3.3 Formulation 213
    10.3.4 Solution 215
10.4 Numerical Approach 219
10.5 Results 219
10.6 Discussion 224
Acknowledgement 227
References 227

11 Modelling of Quasi-Static and Dynamic Load Responses of Filled Viscoelastic Materials **229**
11.1 Introduction 229
11.2 Nonlinear Extension Models, Experiments and Results 231
    11.2.1 Neo-Hookean Extension Models 231
    11.2.2 Approximation of Nonlinear Constitutive Laws 235
11.3 Nonlinear and Hysteretic Models, Experiments and Results 240
    11.3.1 Quasi-Static Hysteresis Loops 240
    11.3.2 A Dynamic Model with Hysteresis 244
11.4 Conclusion 248
Acknowledgements 250
References 250

12 A Gasdynamic–Acoustic Model of a Bird Scare Gun **253**
12.1 Introduction 253
12.2 Model 255
    12.2.1 Geometry 255
    12.2.2 Pot 255
    12.2.3 Jet 258
    12.2.4 Pipe 259
    12.2.5 Radiated Field 260
    12.2.6 Nonlinear Correction in the Pipe 262
12.3 Analysis 264

12.4   Results   266
12.5   Conclusions and Suggestions for Further Work   266
Acknowledgements   268
References   268

13 Paper Tension Variations in a Printing Press   **270**
Preface   270
13.1   Problem Definition   271
13.2   Printing Presses   272
13.3   Modelling   274
       13.3.1   Motion over a Roller   275
       13.3.2   Motion in a Span   281
13.4   The N-Roller Start-up Problem   282
13.5   Concluding Remarks   285
Acknowledgements   288
References   289

Index   **291**

Contributors

Ellis Cumberbatch
Claremont Graduate University, Claremont, CA 91711, USA
ellis.cumberbatch@cgu.edu

Alistair D. Fitt
Faculty of Mathematical Studies, University of Southampton, Southampton
SO17 1BJ, UK
adf@maths.soton.ac.uk

Peter D. Howell
School of Mathematical Sciences, University of Nottingham, University
Park, Nottingham NG7 2RD, UK
Peter.Howell@nottingham.ac.uk

D. Rex Westbrook
Department of Mathematics and Statistics, University of Calgary, Calgary,
Alberta, Canada T2N 1N4
westbroo@ucalgary.ca

Kerry A. Landman
Department of Mathematics and Statistics, University of Melbourne,
Parkville, Victoria 3052, Australia
k.landman@ms.unimelb.edu.au

Mark J. McGuinness
School of Mathematical and Computing Sciences, Victoria University of
Wellington, PO Box 600, Wellington, New Zealand
Mark.McGuinness@vuw.ac.nz

Britta Basse
Department of Mathematics and Statistics, University of Canterbury,
Christchurch, New Zealand
b.basse@math.canterbury.ac.nz

Graeme C. Wake
Department of Mathematics and Statistics, University of Canterbury,
Christchurch, New Zealand
G.Wake@math.canterbury.ac.nz

Donald W. Schwendeman
Department of Mathematical Sciences, Rensselaer Polytechnic Institute,
Troy, 12180-3590, USA
schwed@rpi.edu

Andrew C. Fowler
OCIAM, Mathematical Institute, 24–29, St. Giles, Oxford OX1 3LB, UK
fowler@maths.ox.ac.uk

Helmut Neunzert
Institute for Industrial Mathematics (ITWM), Erwin-Schrödinger-Straße,
D-67663 Kaiserslautern, Germany
neunzert@itwm.uni-kl.de

Norbert Siedow
Institute for Industrial Mathematics (ITWM), Erwin-Schrödinger-Straße,
D-67663 Kaiserslautern, Germany
siedow@itwm.uni-kl.de

Frank Zingsheim
Institute for Industrial Mathematics (ITWM), Erwin-Schrödinger-Straße,
D-67663 Kaiserslautern, Germany
zingsheim@itwm.uni-kl.de

Arnon Chait
Computational Microgravity Laboratory, NASA Glenn Research Center,
Cleveland, OH 44135, USA
Arnon.Chait@grc.nasa.gov

Elizabeth Gray
Computational Microgravity Laboratory, NASA Glenn Research Center,
Cleveland, OH 44135, USA
lizabeth@cml-mail.grc.nasa.gov

Gerald W. Young
Department of Mathematics and Computer Science, The University of
Akron, Akron, OH 44325-4002, USA
gwyoung@uakron.edu

H. T. Banks
Center for Research in Scientific Computation, North Carolina State
University, Raleigh, NC, USA
htbanks@eos.ncsu.edu

Gabriella A. Pintér
Center for Research in Scientific Computation, North Carolina State
University, Raleigh, NC, USA
gapinter@unity.ncsu.edu

Laura K. Potter
Center for Research in Scientific Computation, North Carolina State
University, Raleigh, NC, USA
lkpotter@unity.ncsu.edu

Michael J. Gaitens
Thomas Lord Research Center, Lord Corporation, Cary, NC, USA
info@lordcorp.com

Lynn C. Yanyo
Thomas Lord Research Center, Lord Corporation, Cary, NC, USA
info@lordcorp.com

Sjoerd W. Rienstra
Department of Mathematics and Computing Science, Eindhoven University
of Technology, PO Box 513, 5600 MB Eindhoven, Netherlands
s.w.rienstra@tue.nl

Colin P. Please
Faculty of Mathematical Studies, University of Southampton, Southampton
SO17 1BJ, UK
cpp@maths.soton.ac.uk

Preface

This is a book about *real* industrial problems from *real* industries. The industries and the problems come from all parts of the globe and cover a wide variety of areas, so do the contributors.

We had two specific reasons for producing this book.

First, we believe that applied mathematics is a multifaceted discipline. One of the facets that has emerged and assumed an increasingly important role in the last 30 years has been the need of industrial scientists to use and extend results found by academics. At the same time, academics have become increasingly aware of new phenomena that are of interest to industry. These cross-fertilisations can lead to new and interesting mathematics. This book illustrates the process of industrial/academic interaction through case studies.

Second, we believe that many classical applied mathematics texts treat applications merely as platforms upon which to build mathematical structures. One unfortunate result of this can be that the original requirements which inspired development of the mathematics become hidden. Recently, in an effort to interest students in applied mathematics, and to expose them to the totality of the applied mathematics process, the mathematical modelling process itself (which precedes analysis and numerical considerations) is being taught. This book provides a set of real examples which could be useful for a high-level modelling course.

Introduction

"Technology transfer" has become one of the most well-used phrases of the end of the millennium. The realisation that the worlds of industry and academia cannot fruitfully progress separately has inspired both communities to build strong and mutually beneficial relationships. Often this has meant industry hiring individual professors as consultants, or industry supporting post-doctoral fellows (common in chemistry). An alternative structure has been the utilisation of a quasi-governmental organisation as a go-between, such as NACA/NASA in the US and the Aeronautical Research Council in the UK for work in aeronautics. **Direct** contact between **industry** and the **mathematics community** is more recent, achieving recognition via degree programmes, math-in-industry conferences and journals, all now in a global context. Half of the chapters in this book are products of Study Groups with Industry and Math Clinics, and to some extent the other half derive from similar direct interactions with industry prompted by the successes of these two initiatives.

Study Groups with Industry started in Oxford in 1968 when a small group of applied mathematicians (led by Alan Tayler and Leslie Fox) spent a week problem solving at Oxford University in conjunction with invited representatives from industry. A similar meeting has been held every year since. What *has* changed in recent years is the global nature of this industrial/academic collaboration. In the first 20 or so years Study Groups with Industry only happened in Oxford, and only once a year: by 1999 meetings running on similar lines to the Oxford model had also taken place in Australia, Canada, Denmark, Holland, Indonesia, Mexico, Norway, USA and other locations, in the same year. In many of these ventures domestic applied mathematics societies have encouraged and supported this expansion: pivotal roles have been played in Europe by ECMI (the European Consortium for Mathematics in Industry), in the USA and Canada by SIAM (Society for Industrial and Applied Mathematics) and PIMS (Pacific Institute for the Mathematical Sciences), in Mexico by SMM (Sociedad Matematica Mexicana) and in

1

Australia by ANZIAM (Australian and New Zealand Industrial and Applied Mathematics).

The *modus operandi* of Study Groups with Industry is now well established: ahead of the meeting there is solicitation from industry for the submission of problems, and probably some discussion regarding appropriateness for a week of brain storming. On the first day of the meeting, the academics, their graduate students, and representatives from industry gather. After each industrialist has described their specific unsolved problem, a room is allotted to each industry, and informal groups are formed. The next three days are spent working intensively on each problem, guided and assisted by the representative from industry. Fierce debate often rages: theories and countertheories come and go, and blackboards are filled with equations. Because of the time constraints that apply, the evenings are often used to work on the problems. Normally, a consensus is eventually reached. On the final day of the meeting, progress on each of the problems is summarised and subsequently a technical problem report is published.

Math Clinics originated at the Claremont Colleges in 1973. The structure here has a longer time frame: problems are solicited over the summer, and work on them is contracted for an academic year. Each problem is addressed over this time period by a team consisting of a faculty supervisor and 4–6 graduate and undergraduate students (one of whom is team leader, a managerial position). The team has the time to do basic research, and during the year makes a number of oral and written reports. Students are assigned grades as for a traditional course.

In both the Study Group and Clinic operations, the involvement of students is important. How problems originate, how industrialists expect them to be modelled, and what they expect from an answer, is valuable experience for students aiming for industrial employment, or starting on research in applied mathematics. Additionally students working on these problems learn the dynamics of group work, and the importance of good oral and writing skills.

What type of problems appear at Study Groups and Clinics? This question is difficult to answer, for the range is huge. Not only do the problems vary enormously in physical and mathematical terms, but the reasons that the industrialist has for bringing a specific problem may vary. Typical "industry questions" all begin "we have a process. . . " but may have a range of endings such as

- "... which is well-understood and has worked well for years, but we suspect that it could be made more efficient: how can we optimise it?"

- "... which has worked well for years, but we don't have much idea why. Now we're trying to extend and change it. How does it work?"
- "... which normally works pretty well. Sometimes, though, something goes wrong. Under what circumstances might we expect this to happen, what should we measure to give us warning signs and how might we cure the problem?"
- "... in mind which is new and very promising. Before running expensive and/or dangerous experiments, we'd like to get an idea of how things might turn out."
- "... which works well and we know is safe. But we have to satisfy an external regulatory body to *prove* that it is safe. Can you model the 'worst case' circumstances for us?"
- "... which we have simulated (ourselves, or bought software for). The data and the simulations do not agree. What is going wrong?"

As far as the representatives from industry are concerned, finding the answers to questions of this sort is usually the main priority. There are many other possible advantages of the interaction. The opportunity to spend time discussing their problem with a group of experts is attractive, as is the chance to see what sort of problems other industries have and how they cope with them. Frequently the discussion widens, and other problems may be considered. Finally, there is always the chance to build more longer lasting professional relationships and to recruit promising students.

For the process to be mutually beneficial, everybody concerned must have a good reason for wanting to take part. Some of the benefits for students have been referred to above. What about faculty? It turns out that there are many possible reasons for participating. These include:

- A constant supply of interesting and novel problems which often lead to publications in leading journals
- The chance to form closer relationships with companies which may lead to joint studentships and research contracts
- An opportunity to broaden the range of new problem areas leading to a freshening of the teaching syllabus
- The transfer of mathematics across applications
- A chance to work as part of a team – frequently a novelty for mathematicians.

How much does all of this cost? Considering the possible benefits to both industry and the academics involved, the sums involved are surprisingly small. British Study Groups cost between 10,000–15,000 pounds to run; the money is provided by the participating industries (each problem presenter is asked to contribute about 1000 pounds), and various other organisations such as the London Mathematical Society and (until recently) EPSRC also contribute much-needed revenue. As far as the industrial participants are concerned, the biggest cost is normally that of having their expert in a particular field away from the office for a whole week. For clinics, the financial details are rather more large-scale; each clinic project costs the industry concerned a sum between 35,000 and 45,000 dollars. Although this might sound a lot, for their money the client gets a year's work from a dedicated group of experts: the productivity often equals that of an engineer-year which costs the employer five times as much. Do they consider this money well spent? This is a question that can only truly be answered by the industrial scientists themselves, but it is significant that a large number of projects have been brought to the Claremont clinics by "repeat customers".

Thirteen different problems are considered in this book; all originate from real collaborations with industry. Although some of them are a little difficult to classify, four of the problems are recognisably elliptic, six are parabolic, and two are hyperbolic. This distribution may be said to fairly represent the frequency of each type of problem that is normally encountered at Study Groups.

The order chosen for the articles is related to ordinary and partial differential equation classification. Chapters 1–4 concern models based on essentially elliptic partial differential equations. Chapters 5–10 all involve some form of diffusion, with nonlinear effects, convection and reaction, and finally radiation being successively introduced. The classification of the underlying equations in chapter 11 is less clear; the equations contain both parabolic and hyperbolic features. The concluding two chapters, 12 and 13, address two hyperbolic problems.

0

Mathematical Preliminaries

0.1 The Continuum Model

Most of the modelling introduced in the following chapters uses the continuum approach. In this introductory chapter, we list the commonly used equations: those describing diffusion, convection, radiation, and fluid and solid mechanics. We do not attempt to give an even partially rigorous derivation of any of these equations; our purpose is to provide a ready resource, and to indicate source books of wider scope. Above all, this section should be seen as answering the question "why did they start from *those* equations?".

Let us approach the continuum model by considering the example of diffusion. Diffusion is a molecular process. Consider the diffusion of heat: the diffusion happens because a hot region of a material has molecules of higher energy than those in cooler parts. Energy equalisation therefore takes place by molecular interaction – and the heat is said to "diffuse". To enable us to view this at the continuum level, local averages (for example, over many molecules) are taken: the molecular picture is smeared. The concept of heat as a variable having a value only at molecular sites is replaced by a framework in which heat is regarded as a variable that has continuous values. The laws governing changes in the continuous functions to describe heat transfer are treated "phenomenologically". Models are created at both levels; their usefulness, success, and relevance depends on the application.

0.1.1 Conservation Laws

Physical phenomena expressed in the continuum paradigm are usually phrased in terms of conservation laws. Let us introduce this in a generic fashion: we consider a substance, A, distributed continuously. For a subset V of the continuum,

$$\int_V A \, dV$$

5

is the amount of our substance contained in V, so that $A(x)$ measures the local amount of A per unit volume at location x at time t. We anticipate that A is changing with time and with location, within (and outside) V, by its own motion relative to the continuum and by the motion of the continuum carrying it around. Let Q denote the flux of A, defined so that the amount of our substance flowing per unit time across the surface S surrounding V is

$$\int_S Q \cdot n \, dS,$$

where n is the unit outward normal to S. This definition means that, locally, for a small, flat element of area dS and normal n, the amount of substance flowing across dS per unit time is $Q \cdot n \, dS$. We now assert that the rate of change of the amount of substance within V is due only to the amount flowing across its boundary. (That is, we assume no sources or sinks of A within V.) This gives

$$\frac{\partial}{\partial t} \int_V A \, dV = - \int_S Q \cdot n \, dS. \tag{0.1}$$

The right-hand side is transformed to a volume integral by use of the divergence theorem. Also, for a volume V fixed in space, the time derivative may be taken inside the integral. Thus (0.1) becomes

$$\int_V \left(\frac{\partial A}{\partial t} + \nabla \cdot Q \right) dV = 0. \tag{0.2}$$

Since V is an arbitrary volume, it follows that the integrand is identically zero throughout the continuum. This argument relies on the assumption that the integrand is continuous, and it fails in its absence. (The latter occurs when shock waves are present.) This gives the *local* equation

$$\frac{\partial A}{\partial t} + \nabla \cdot Q = 0. \tag{0.3}$$

This equation is evident in various forms as we quote equations of continuum mechanics below.

0.2 Diffusion

The phenomenon of diffusion is all around us. We drink hot tea, and its heat, concentrated at first, is diffused throughout the body. The process is helped by warmed blood flowing to the colder extremities – this is advection. Similar processes are happening everywhere – concentrations of heat, moisture, salt, smoke, and other pollutants, are being dispersed in the atmosphere and oceans by the processes of diffusion and advection.

Let us consider pure diffusion first, and use the diffusion of heat as an illustrative example. The principle of the conservation of heat induces us to define the quantity A in (0.1) to be

$$A = \rho c_p T \tag{0.4}$$

where ρ is the mass density of the material in which the heat resides, c_p is the specific heat of the material, and T is its temperature. The flux of heat across a surface is taken as

$$Q = -k\nabla T \tag{0.5}$$

where the factor k is called the thermal conductivity. This model, which asserts that the rate of heat flow is proportional to the temperature gradient (heat flows from high temperatures to colder ones) seems a natural one – and it works wonderfully well! With the substitutions of A and Q from (0.4) and (0.5) into the general conservation law (0.3), we find that

$$\frac{\partial T}{\partial t} = D\nabla^2 T \tag{0.6}$$

where

$$D = \frac{k}{\rho c_p} \tag{0.7}$$

is called the thermal diffusivity. In obtaining this equation, we have assumed that the component factors that make up D are all constant. This assumption may not hold in various applications, in which case T is governed by a more complicated equation than (0.6).

In general, let us consider the variable A in (0.1) to be the concentration $c(x, t)$ of a substance which is diffusing. The flux law for the diffusion of c similar to (0.5) is given by $Q = -D\nabla C$. This is known as Fick's law. This alone would give (0.6) as the conservation law, with c replacing T. However, let us now add the process of advection to Fick's law. In addition to the concentration gradient ∇c transferring substance within the material, there is also material flow. If $u(x, t)$ denotes the velocity of the medium, the material crossing a surface element will contribute a term $cu \cdot n$ to the rate of transfer of c. Hence Fick's law when diffusion and advection are both present reads

$$Q = -D\nabla c + cu \tag{0.8}$$

and (0.3) becomes

$$\frac{\partial c}{\partial t} + \nabla \cdot (c\boldsymbol{u}) = D\nabla^2 c. \tag{0.9}$$

Often the material velocity $\boldsymbol{u}$ may be taken as constant, in which case (0.9) simplifies. In general, however, $\boldsymbol{u}$ satisfies other equations which must be coupled with (0.9).

The equations of motion require boundary conditions. Let us now discuss these with relation to the heat diffusion equation (0.6); extension to the general diffusion equation (0.9) is straightforward. On a surface S at which the temperature is constant or is specified, we have

$$T = \text{constant} \quad \text{or} \quad T = F(\boldsymbol{x}, t) \tag{0.10}$$

for $\boldsymbol{x} \in S$.

On surfaces S that are insulated or across which the flux of heat is specified, we have

$$\frac{\partial T}{\partial n} = 0 \quad \text{or} \quad \frac{\partial T}{\partial n} = G(\boldsymbol{x}, t) \tag{0.11}$$

where $\boldsymbol{n}$ is the unit normal to the surface S and again $\boldsymbol{x} \in S$. There are situations for which neither (0.10) nor (0.11) apply, and a mixed boundary condition

$$\frac{\partial T}{\partial n} = \alpha(T - T_0) \tag{0.12}$$

is more appropriate for $\boldsymbol{x} \in S$. This boundary condition relates the flux of heat transferred across a boundary to the difference between the boundary temperature and the ambient temperature. It is often used at a solid–fluid interface where the fluid may be in motion. The coefficient α is an empirical constant, varying with the situation encountered. This boundary condition is often called the "radiation" boundary condition, though it is not related to radiative heat transfer.

The mathematical designation for (0.10), (0.11) and (0.12) are Dirichlet, Neumann, and Robin (or mixed) boundary conditions, respectively: for extensive discussion of the heat diffusion equation, its properties, and its engineering applications, the reader is referred to [4], [5], [11] and [22].

A boundary condition met in many applications concerns change of phase at an interface. In the case of the heat equation, the conditions are then known as the Stefan boundary conditions. In particular, consider modelling the melting of ice and, for illustrative purposes, take a plane geometry. Ice, at temperature $T = 0°C$, initially occupies the half-space $x > 0$. Beginning at $t = 0$ the ice is melted by applying a heat source at $x = 0$ which maintains a temperature $T = T_o > 0$ there. At time $t > 0$ there is water in $0 < x < s(t)$ and ice in $x > s(t)$, where $x = s(t)$ denotes the moving interface that separates ice from water. The Stefan boundary conditions at this interface are

$$T = 0 \quad \text{and} \quad -k\frac{\partial T}{\partial x} = L\rho\frac{ds}{dt} \quad \text{at} \quad x = s(t). \tag{0.13}$$

The boundary condition (0.13) is to be used in finding the temperature profile in $0 < x < s(t)$. In (0.13) L is the coefficient of latent heat, that is, the amount of heat per unit mass required to change ice into water at $T = 0^oC$. Equation (0.13) gives two equations to be satisfied at $x = s(t)$, as distinct to the single condition in (0.10), (0.11), or (0.12) prescribed previously. Since the location of the ice–water interface is not known *a priori*, two conditions are necessary to solve for the temperature profile and the location $x = s(t)$ simultaneously. Problems of this type are known as free or moving boundary value problems; they are generally difficult to solve as they are nonlinear. For further details the reader is referred to [8] and [21].

0.3 Flow and Deformation of Solids and Fluids

In many of the problems in this book, a fluid or a solid deforms under the action of external forces; to solve the problem under consideration it is necessary to determine how the fluid or solid moves. We therefore need to consider how to propose general equations for the flow of such materials.

We distinguish between four states of matter: solid, fluid, gas, and plasma. Determining exactly which of these four states of matter best describes a given piece of matter is a highly nontrivial process, and we do not intend to pursue this in a detailed manner; rather we content ourselves with simple "coffee table" explanations. Consider a bucket of matter. If the bucket contains a solid, then the constituent particles of the solid are disinclined to move relative to each other. Moreover, if the bucket is "poured out", then the solid will retain its shape. If the bucket contains a fluid, then particles can move relative to each other with much greater ease. Inversion of the bucket will now result in a differently shaped fluid. If the bucket contains a gas, then the gas will not retain its "shape" even *before* the bucket is poured out. Our approach is thus broadly as follows: we acknowledge that a plasma state of matter exists, but give no attention to modelling it, and we treat gases as "fluids" since the constituent particles of both gases and fluids can move relative to each other with ease.

0.3.1 Equations of Motion

To model the flow of solids and fluids, we shall assume that the rate of change of momentum is equal to the sum of the applied forces; this is Newton's second law.

The applied forces separate into two categories. There are body forces *b* acting on each continuum *volume* element: the force per unit volume is ρb, where ρ is the mass density. The simplest body force (and the only one, for most of non-electromagnetic physics) is that due to gravity. The second category of force for a continuum is that acting on *surface* elements. This is manifested by considering simple actions: (1) blowing up a balloon so that increasing the interior pressure expands the balloon volume until it equalises with the exterior ambient pressure. These pressures are expressed in terms of force per unit surface area, and elementary physics shows that in the static case the pressure acts in a direction normal to surface elements; (2) rotating a solid body, say a cylinder, immersed successively in two different fluids requires different torques to achieve the same angular velocity – the "stickier" (more viscous) the fluid the higher the torque required. It is evident here that there are surface forces acting tangential to the surface of the moving cylinder.

In general, the surface force acting on a surface with normal *n* has both normal and tangential components, and this force will depend on the orientation of *n*. Again, elementary physical principles generate expressions for the force per unit area, *F*, in terms of the normal *n*: since *F* and *n* each have three components, the magnitude factor must have nine components, as in matrix multiplication. In terms of components,

$$F_i = \sum_{j=1}^{3} T_{ij} n_j$$

where

$$T = \{T_{ij}\} \tag{0.14}$$

is the stress tensor or matrix.

A conservation law for linear momentum similar to (0.1) may now be invoked. Since the surface forces (0.14) may be expressed as a product of the stress and the normal vector, use of the divergence theorem allows the surface integral to be expressed in terms of a volume integral, as in (0.2). The local equation for conservation of linear momentum is therefore derived, similar to (0.3), as

$$\rho a = \text{div } T + \rho b. \tag{0.15}$$

Although (0.15) is the basic conservation of momentum equation for continuum mechanics, much remains to be done before it is of practical use. We note immediately that, even if we assume that the body force b is known, the stress tensor T has nine components and the acceleration a has three. Thus (0.15) comprises three equations in twelve unknowns.

Further physical principles may be invoked to simplify (0.15). We expect that angular as well as linear momentum will be conserved, and it transpires (see, for example [20]) that this amounts to the fact that the stress tensor is symmetric, so that $T = T^T$. To simplify things further, we have to be more specific about what sort of material we are dealing with. This is done by specifying the "constitutive law" of the material: this law delineates the stress in terms of kinematic (and possibly other) quantities. Constitutive laws cannot be arbitrarily specified, since possible forms of the stress tensor are actually quite severely limited by "frame indifference" (the principle that the results of any experiment should be independent of the frame of reference used to make measurements), and other similar constraints. Nevertheless, experiments show that different types of fluid move in different ways, and theories of more or less complexity may be proposed to describe some of the more exotic materials; below, we shall content ourselves with delineating only the simplest types of flow and deformation.

The fact that we cannot proceed further until a constitutive law has been specified means that at this point our continuum theory begins to separate along different branches; we shall examine these one by one.

0.4 Incompressible Linear Viscous Fluid

First, consider a viscous fluid. We assume that the fluid has dynamic viscosity μ and that its motion depends linearly on derivatives of the infinitesimal rate of strain (i.e. spatial derivatives of the fluid velocities). We also assume that the stress tensor T depends linearly on the velocity gradients. Once we have decided this, it is fairly easy to show (see, for example [20]) that

$$T_{ij} = -p\delta_{ij} + 2\mu d_{ij} \tag{0.16}$$

where p denotes the pressure, δ_{ij} is the "Kronecker delta" satisfying $\delta_{ij} = 0$ when $i \neq j$ and $\delta_{ij} = 1$ when $i = j$, and d_{ij} is the "infinitesimal rate-of-strain tensor" given by

$$d_{ij} = \frac{1}{2} \left(\frac{\partial u_i}{\partial x_j} + \frac{\partial u_j}{\partial x_i} \right). \tag{0.17}$$

Two matters remain to be taken care of before we can produce a closed system of equations from (0.15), for we need (i) to express the acceleration vector a in terms of the velocity and (ii) to ensure that the fluid is incompressible and that mass is conserved. The first of these is accomplished by noting that the acceleration should be thought of as the rate of change of the velocity *following the fluid* (since this is derived from Newton's second law, which states that "at a *particle*, the forces. . . "). Thus (for fuller details, see [1])

$$a = \frac{Du}{Dt} \equiv u_t + (u \cdot \nabla)u \tag{0.18}$$

where the velocity u is (u_1, u_2, u_3). The second term on the right-hand side of (0.18) is nonlinear, and presents the first of many hurdles that exist as far as simple analysis of the equations is concerned.

The conservation of mass equation may be derived by considering a "control volume" as was done in section 0.1.1; when combined with incompressibility ($\rho = $ constant), we find that

$$\nabla \cdot u = 0.$$

We now use (0.16) and (0.17) in (0.15) to give

$$\nabla \cdot u = 0$$

$$u_t + (u \cdot \nabla)u = -\frac{1}{\rho}\nabla p + \nu\nabla^2 u + b \tag{0.19}$$

where $\nu = \mu/\rho$ is the kinematic viscosity.

These are the famous Navier–Stokes equations for the flow of an incompressible linear viscous fluid; four equations must be solved for the three components of the velocity u and the pressure p. Many of the industrial problems considered in this book begin with these equations. A large number of books have been written on the theoretical properties of these equations (see, for example [14], [25], and [26]), and we do not propose to discuss them in any detail; for the moment, we simply note that

- The equations require suitable initial and boundary conditions; these consist of specifying $p(x)$ and $u(x)$ at $t = 0$ and asserting that, at any solid boundary in the flow, the flow velocity is identical to that of the boundary (and thus, for stationary boundaries, $u = 0$ – the famous "no-slip" condition).
- The equations are nonlinear, and closed-form solutions are very hard to obtain for all but the simplest of flows.

- Because of the rarity of exact solutions, simplified versions of the Navier–Stokes equations are often used. These are approximations, valid in various flow regimes.

- The pressure p plays the role of a Lagrange multiplier in the equations; this has the effect of making the numerical solution of the equations a highly nontrivial task.

- An alternative formulation of the problem may be obtained by taking the curl of (0.19) to eliminate p. The resulting equation may be thought of as an evolution equation for the vorticity $\omega = \mathrm{curl}\ u$.

- The nonlinear nature of the equations suggests that both laminar and turbulent flows satisfy the Navier–Stokes equations; in the latter case, solutions may be expected to be chaotic.

We now look at various specialisations and generalisations of the Navier–Stokes equations that form the starting point for many of the models considered in this book.

0.4.1 Inviscid Incompressible Fluid

For inviscid flow, we set $\nu = 0$ in (0.19) and obtain

$$\nabla \cdot u = 0$$

$$u_t + (u \cdot \nabla)u = -\frac{1}{\rho}\nabla p + b, \tag{0.20}$$

which are usually referred to as the Euler equations. Although much simpler than the full Navier–Stokes equations, the Euler equations are still nonlinear, and so exact solutions are still hard to find. As a system, they are of lower order than the Navier–Stokes equations, and so a full no-slip boundary condition cannot be applied, and we can insist only that there is no flow normal to a solid boundary. This is a severe deficiency in the equations, which may lead to paradoxical results (see, for example [2]).

Note that the body force b in (0.20), when conservative, may be written as a gradient, in which case it may be absorbed by redefining the pressure. When the body force is gravity, p is replaced by $p + \rho g z$, where z is the vertical coordinate. For steady flow, and a general conservative force with potential χ, (0.20) may be rewritten using standard vector identities as

$$\nabla\left(\frac{1}{2}u^2\right) - u \wedge (\nabla \wedge u) = -\frac{1}{\rho}\nabla p - \nabla\chi.$$

Rearranging this gives

$$\boldsymbol{u} \wedge \boldsymbol{\omega} = \nabla \left(\frac{1}{2} u^2 + \frac{p}{\rho} + \chi \right)$$

and hence, we derive the famous Bernoulli law, which states that for steady inviscid incompressible flow the quantity

$$\frac{1}{2} u^2 + \frac{p}{\rho} + \chi$$

is constant along a streamline or a vortex line.

0.4.2 Inviscid Incompressible Irrotational Flow

For irrotational flow with $\boldsymbol{\omega} = \text{curl } \boldsymbol{u} = \boldsymbol{0}$, huge simplifications are possible. By a standard theorem the fact that curl $\boldsymbol{u} = \boldsymbol{0}$ guarantees the existence of a scalar potential (the "velocity potential") ϕ such that $\boldsymbol{u} = \nabla \phi$. Instead of having to determine three components of velocity, the problem has thus been reduced to that of finding a single scalar potential. Moreover, the equation satisfied by ϕ is particularly simple: by using conservation of mass $\nabla \cdot \boldsymbol{u} = 0$, we find that

$$\nabla^2 \phi = 0. \tag{0.21}$$

For two-dimensional flows, standard complex-variable methods may now be applied to solve for the velocity potential in a wide variety of cases (see, for example [17]). Extensive knowledge exists regarding the solutions and mathematical properties of (0.21), and much progress can often be made. The price that has to be paid for this progress, of course, is that the full no-slip condition cannot be applied, and so drag predictions are not possible, and the details of the flow near to solid bodies are not correct.

For inviscid incompressible irrotational (but unsteady) flow, the momentum equations may also be integrated to give another version of the Bernoulli equation, which states that the quantity

$$\phi_t + \frac{p}{\rho} + \frac{1}{2} \mid \nabla \phi \mid^2 \tag{0.22}$$

is a function of time alone; this may be regarded as an equation to determine the pressure p once (0.21) has been solved to determine ϕ, and hence $\boldsymbol{u}$.

0.4.3 Compressible Viscous Flow

When a fluid is both viscous and capable of compression, matters become altogether more complicated. The density ρ is now unknown and must be determined as part of the solution. The conservation of mass equation now involves derivatives of ρ and, since $\nabla.\boldsymbol{u}$ is nonzero, the momentum equations are further complicated. The Navier–Stokes equations thus become

$$\rho_t + \nabla \cdot (\rho \boldsymbol{u}) = 0 \tag{0.23}$$

$$\boldsymbol{u}_t + (\boldsymbol{u} \cdot \nabla)\boldsymbol{u} = -\frac{1}{\rho}\nabla p + \nu\nabla^2\boldsymbol{u} + (\zeta + \nu/3)\nabla(\nabla \cdot \boldsymbol{u}) \tag{0.24}$$

where ζ is usually referred to as the "bulk" viscosity. Another equation is required to close the system, and this is usually taken to be some simple relationship between the pressure, density, and temperature. The "perfect gas" law $p = \rho R T$, where R is the universal gas constant and T is the temperature, is a typical example, but of course this introduces yet another new unknown, T; so, unless this is assumed to be constant, an energy equation is required to close the system.

0.4.4 Compressible Inviscid Flow

Because of the difficulties associated with analysing compressible viscous flow, it is often assumed that the fluid concerned is inviscid. Since the dynamic viscosities of gases tend to be a great deal lower than those of fluids, and gases tend to be compressible while most fluids are largely incompressible, the assumption of inviscid flow is frequently very accurate as far as gases are concerned. This avoids the complication of calculating the work done by the viscous stresses, and the consequent changes in the heat flow and their effects on the thermodynamics. Even so, the latter is still formidable in the most general context, and here we introduce the simplest approximation (which is accurate in many circumstances) and refer the reader to more comprehensive texts (see for example [9] and [23]) for further guidance. This approximation is valid for applications (for example acoustics) in which expansion and compression of the gas take place so rapidly that the heat loss and gain do not have time to be conducted to neighbouring fluid particles. These assumptions can be shown to imply that

$$p = K\rho^\gamma \tag{0.25}$$

where γ is the ratio of the specific heats at constant pressure and constant volume respectively (thus $\gamma = c_p/c_v \sim 7/5$ for air).

For a compressible fluid such that the pressure is a function only of density, as in (0.25), the Bernoulli equation (0.22) generalises to state that the quantity

$$\frac{\partial \phi}{\partial t} + \int \frac{dp}{\rho} + \frac{1}{2}|\boldsymbol{u}|^2 \tag{0.26}$$

is a function of time alone. Equations (0.23), (0.25), and (0.26) now provide
the basis for compressible inviscid irrotational fluid mechanics.

These equations are nonlinear, and closed-form results are hard to obtain.
Some exact solutions, including flows exhibiting shock formation, are available for plane time-dependent flows (see, for example [9]). Linear equations
govern perturbations about a uniform state. For a base flow with uniform
velocity $u = Ue_x$ and density ρ_0, the perturbation velocity satisfies

$$\nabla^2\phi - M^2\phi_{xx} = \phi_{tt} \tag{0.27}$$

where

$$M = \frac{U}{a_0}, \quad a_0^2 = \left(\frac{dp}{d\rho}\right)_{\rho=\rho_0}.$$

Here M is the Mach number: the ratio of the base flow speed U to a_0, the
sound speed in the ambient fluid. For $M < 1$, (0.27) describes subsonic
aerodynamics, and for $M > 1$ it describes supersonic flow. For $M = 0$, (0.27)
reduces to the standard wave equation which describes the propagation of
small disturbances in an otherwise stationary fluid.

0.4.5 The Reynolds Number and Flow in a Viscous Boundary Layer

The subject of viscous flow contains a dilemma: should we use the full Navier–
Stokes equations, which are too complicated to solve in all but the simplest
circumstances, or the Euler inviscid equations, which do not allow a full
prescription of boundary conditions? It turns out that, to examine viscous
effects close to a solid boundary, a wise approach is to decide which terms in
the equations are important by carefully analysing the sizes of each of them.
This process is most easily carried out by nondimensionalising the equations,
a process that takes place in many of the studies contained in this book. If we
assume that "typical" lengths and velocities in the flow are given by L and U
respectively, and write $t = (L/U)t'$, $x = Lx'$, $u = Uu'$, and $p = \rho U^2 p'$,
then the Navier–Stokes equations become (after dropping the primes)

$$\nabla \cdot u = 0$$

$$u_t + (u \cdot \nabla)u = -\nabla p + \frac{1}{\text{Re}}\nabla^2 u$$

where the Reynolds number Re is defined by

$$\text{Re} = \frac{LU}{\nu}.$$

When the Reynolds number is very large (more than a few thousand, say) experiments show that the flow is turbulent. For values of Re that are much in excess of unity but still small enough for the flow to be laminar, a careful analysis shows that the flow is essentially inviscid unless we are within a distance of order $L/\sqrt{\text{Re}}$ of a solid boundary; in this region, the "boundary layer", viscous effects are important and cannot be ignored. The length scales parallel and transverse to the boundary layer are in the ratio $1/\sqrt{\text{Re}}$. Introducing the relevant scalings to reflect this, and estimating appropriate sizes for the velocity components, allows the Navier–Stokes equations to be simplified (see for example [20]). The boundary layer equations for flow along a flat plate with a constant external velocity U turn out to be

$$uu_x + vu_y = \nu u_{yy}$$

$$u_x + v_y = 0$$

and the boundary conditions are $u = v = 0$ on $y = 0$ and the "matching" condition $u \to U$ as $y \to \infty$.

The boundary layer concept introduced by Prandtl in 1904 was found to be very efficient in obtaining the drag forces on slender bodies such as aerofoils. The theory has been extended and generalised to cover many other situations in which substantial changes occur across thin layers: these include shock waves, shell theory, and chemical reactions among many others. The boundary layer concept is now part of a general mathematical theory called MAE (Matched Asymptotic Expansions), and good treatments of this material may be found in [12], [13], [18], [19], and [27].

0.4.6 Viscous Flow at Low Reynolds Number

We saw above that simplifications are possible when the Reynolds number is much greater than one; the equations also become much more amenable to analysis when Re is much smaller than one. (Actually this approximation frequently gives good results even when Re $\sim$ 1.) Carrying out the same nondimensionalisation as before, save for the fact that now we scale the pressure using $p = (\mu U/L)p'$ (the relevant scaling for viscosity-dominated flow), we find (after again dropping the primes) that the nondimensional equations become

$$\nabla \cdot \boldsymbol{u} = 0$$

$$\text{Re}(\boldsymbol{u}_t + (\boldsymbol{u} \cdot \nabla)\boldsymbol{u}) = -\nabla p + \nabla^2 \boldsymbol{u}.$$

Since the small parameter Re does not now multiply the highest derivative
in the equation, the problem may be solved by simple regular perturbations.
Proceeding to the limit Re $= 0$, we find that "slow viscous flow" satisfies the
equations

$$\nabla p = \mu \nabla^2 \boldsymbol{u}, \quad \nabla \cdot \boldsymbol{u} = 0.$$

The analysis is thus significantly simplified in this case, for the equations
of motion are linear. For two-dimensional flow it is convenient to eliminate
the pressure by taking the curl of the momentum equation. If we do this and
define a stream function $\psi(x, y)$ such that $u = \psi_y$ and $v = -\psi_x$ (so that the
continuity equation is automatically satisfied), then the slow flow equations
reduce to the biharmonic equation

$$\nabla^4 \psi = 0.$$

Although this does not look much more complicated than the Laplace equation that was recovered for irrotational inviscid incompressible flow, it
does not yield readily to complex-variable methods in nearly such a fruit-
ful manner as the Laplace equation, since conformal maps preserve only
the Laplace and not the biharmonic operator. Valuable progress has nevertheless been made using complex-variable methods (see for example [10]
and [15]).

0.4.7 Viscous Flow in a Thin Layer

When viscous flow takes place in a "thin" layer, yet more simplifications are
possible; we analyse the most basic situation where, for a two-dimensional
viscous flow say, typical y distances are much less than typical x distances.
Characterising these distances by h and L respectively, the obvious small
parameter to exploit is $\epsilon = h/L$. If the Navier–Stokes equations are now
nondimensionalised using $x = Lx'$, $y = \epsilon Ly'$, $u = Uu'$, $v = \epsilon Uv'$ (where
U is a typical horizontal speed in the flow), and we use the "lubrication"
pressure scaling $p = (\mu U)/(L\epsilon^2)p'$, then, dropping the primes, we easily
find that, to leading order in ϵ, the equations of motion become simply

$$p_x = \mu u_{zz}, \quad p_y = 0, \quad u_x + v_y = 0$$

and the fact that p is a function of x alone now allows the first of these
equations to be integrated immediately. An equation for the pressure may
now be found by integrating across the layer: in the simplest circumstances
we retrieve Reynolds' equation

$$(h^3 p_x)_x = 6\mu U h_x$$

where $h(x)$ is the shape of the top boundary. Many variations on this theme are possible, but all rely on the stricture that both ϵ and the "reduced Reynolds' number" ϵ^2 Re are small. For further details, the reader is referred to [1], [3], and [20].

0.5 Linear Elasticity

Suppose that we wish to analyse the motion of a solid rather than a fluid. We know that (0.15) still applies, but what changes should be made to the constitutive laws to reflect the differences between solids and fluids? Once again, many complicated models may be proposed. In particular, many materials possess qualities of both fluids and solids, so that viscoelastic models may be required. For the simplest sort of solid, however, we will invoke Hooke's law that tension is proportional to extension. The stress tensor still needs to satisfy the usual basic physical laws such as symmetry and objectivity, and, using similar reasoning to the viscous fluid flow case, we find that it may be written

$$T_{ij} = \lambda e_{jj}\delta_{ij} + 2\mu e_{ij} \tag{0.28}$$

where e_{ij} denotes the "infinitesimal strain tensor" defined by

$$e_{ij} = \frac{1}{2}\left(\frac{\partial u_i}{\partial x_j} + \frac{\partial u_j}{\partial x_i}\right).$$

The vector u is now the vector of *displacements* (not of velocity components as in previous sections of this chapter), and λ and μ are the Lamé elastic constants, which are related to the more commonly used Young's modulus E and Poisson's ratio v by

$$\lambda = \frac{vE}{(1+v)(1-2v)}, \qquad \mu = \frac{E}{2(1+v)}.$$

Equations of motion may now be derived by using (0.28) in (0.15). We find that

$$(\lambda + \mu)\frac{\partial^2 u_j}{\partial x_i \partial x_j} + \mu\frac{\partial^2 u_i}{\partial x_j \partial x_j} = \rho u_{itt}. \tag{0.29}$$

In these equations there is no equivalent to the pressure in the Navier–Stokes equations; elastic bodies are, in general, compressible. Unlike the Navier–Stokes equations, however, the kinematics used to derive (0.29) means that they are valid only for small displacements – in contrast to the Navier–Stokes equations, which hold for velocities of arbitrary size (see, for example [6] and [7]).

0.6 Summary

We end by summarising our views on the material contained in this chapter and how it is used in the rest of this book:

- Space has only allowed us to give the briefest of summaries of the equations that will form the starting point of many of the problems dealt with in this book. The reader is encouraged to seek out further references to the theory.
- In the process of mathematical modelling, the general idea is to identify the key parameters and mechanisms that control the physical behaviour of the system under investigation, thereby obtaining equations that relate these mechanisms. Although the starting point for this process is very often one or more of the equations that appear in this chapter, much ingenuity is frequently required to derive a final model that is tractable. For information and examples on the general modelling process, the reader is referred to [16], [24], or [28].
- Often the equations that result from the initial modelling process will themselves present formidable challenges. Frequently asymptotic analysis has to be used to simplify the problem. If this is not possible then often numerical methods must be used to elucidate the solution.
- The modelling process is iterative: the results from the initial approach must be compared with data to further inform the modelling process. This may involve frequent model reformulations.

References

[1]   Acheson, D. (1990) *Elementary Fluid Dynamics*, Oxford: Clarendon Press.
[2]   Birkhoff, G. (1950) *Hydrodynamics*, Cincinnati: Princeton University Press.
[3]   Cameron, A. (1983) *Basic Lubrication Theory*, Chichester: Ellis Horwood.
[4]   Cannon, J. R. (1984) *The One-Dimensional Heat Equation*, California: Addison-Wesley.
[5]   Carslaw, H. S. & Jaeger, J. C. (1947) *Conduction of Heat in Solids*, Oxford: Clarendon Press.
[6]   Chadwick, P. (1976) *Continuum Mechanics*, New York: John Wiley.
[7]   Chung, T. J. (1988) *Continuum Mechanics*, New Jersey: Prentice-Hall.
[8]   Crank, J. (1984) *Free and Moving Boundary Problems*, Oxford University Press.
[9]   Courant, R. & Friedrichs, K. O. (1948) *Supersonic Flow and Shock Waves*, New York: Interscience.
[10] England, A. H. (1971) *Complex Variable Methods in Elasticity*, London: Wiley Interscience.
[11] Hill, J. M. & Dewynne, J. N. (1991) *Heat Conduction*, CRC Press.
[12] Hinch, E. J. (1991) *Perturbation Methods*, Cambridge University Press.
[13] Kevorkian, J. & Cole, J. D. (1968). Appl. Math. Sci. **34**, Springer-Verlag.

[14] Ladyzhenskaya, O. A. (1963) *The Mathematical Theory of Viscous Incompressible Flow*, New York: Gordon & Breach.

[15] Langlois, W. E. (1964) *Slow Viscous Flow*, New York: Macmillan.

[16] Lin, C. C. & Segal, L. A. (1974) *Mathematics Applied to Deterministic Problems in the Natural Sciences*, New York: Macmillan.

[17] Milne-Thomson, L. M. (1949) *Theoretical Hydrodynamics*, London: Macmillan.

[18] Murray, J. D. (1984) *Asymptotic Analysis*, New York: Springer.

[19] Nayfeh, A. H. (1973) *Perturbation Methods*, New York: John Wiley.

[20] Ockendon, H. & Ockendon, J. R. (1995) *Viscous Flow*, Cambridge University Press.

[21] Ockendon, J. R. & Hodgkins, W. R. (1975) *Moving Boundary Problems in Heat Flow and Diffusion*, Oxford: Clarendon Press.

[22] Ozisik, M. Ni. (1968) *Boundary Value Problems of Heat Conduction*, Ohio: International Textbook Co.

[23] Stanyukovich, K. P. (1960) *Unsteady Motion of Continuous Media*, Oxford: Pergamon Press.

[24] Tayler, A. B. (1986) *Mathematical Models in Applied Mechanics*, Oxford: Clarendon Press.

[25] Temam, R. (1983) *Navier–Stokes Equations and Nonlinear Functional Analysis*, CBMS-NSF Regional Conference Series, SIAM.

[26] Temam, R. (1984) *Navier–Stokes Equations*, Amsterdam: North-Holland.

[27] Van Dyke, M. (1975) *Perturbation Methods in Fluid Dynamics*, Parabolic Press.

[28] Wan F. Y. M. (1989) *Mathematical Models and their Analysis*, New York: Harper & Row.

1

Fluid-Mechanical Modelling

of the Scroll Compressor

Preface

This case study concerns the flow of gas in a so-called *scroll compressor*. In this device a number of chambers of gas at different temperatures and pressures are separated by narrow channels through which leakage can occur. Using compressible lubrication theory, an estimate for the leakage rate is found in terms of the material properties of the gas and the geometry of the compressor. Thus a simple functional is obtained which allows the efficiency of different compressor designs to be compared. Next we derive a set of ordinary differential equations for the temperature and pressure in each chamber; the coupling between them arises from the leakage. The numerical solution of these equations allows a realistic simulation of a working compressor, and suggests some interesting possibilities for future designs.

This problem arose at the *32nd European Study Group with Industry* held in September 1998 at the Technical University of Denmark – the first to be held outside the United Kingdom. It was presented by Stig Helmer Jørgensen from DANFOSS, which is Denmark's largest industrial group and specialises in controls for refrigeration and heating. The Danish Study Group was a great success and is expected to be repeated annually henceforth. The feedback from DANFOSS has also been encouraging, and hopefully this represents the start of a long-term collaboration.

1.1 Introduction

The scroll compressor is an ingenious machine used for compressing air or refrigerant, which was invented in 1905 by Creux [1]. Unfortunately, technology was insufficiently advanced at the time for workable models to be manufactured, and it was not until the 1970s that commercial interest in the idea was revived – see e.g. [3]. The device consists of two nested identical scrolls, one of which is rotated through $180°$ with respect to the other. In the classical design, both scrolls are *circle involutes* as shown in figure 1.1.

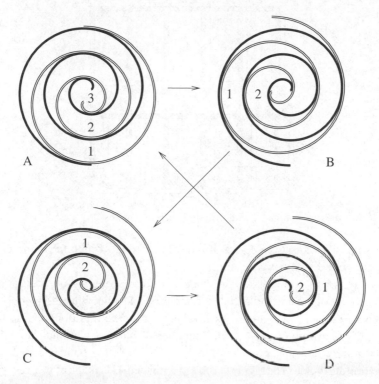

Fig. 1.1. Schematic diagram of a scroll compressor

Now, while the darker scroll is held fixed, the other is moved through a clockwise circular motion such that the two always remain in contact, going successively through the positions shown in diagrams A, B, C, and D of figure 1.1. Each scroll is fitted to a backplate, so that a side view looks something like figure 1.2: the top diagram is for a section in which the scrolls are in contact, and the bottom one represents any other section. Notice the gaps at the sides where gas can be ingested and the outlet gap in the middle.

To see how the compressor works, consider first the pocket of gas marked "1" in diagram A of figure 1.1. This has just been ingested from the gas outside the scrolls and is now sealed inside what we shall henceforth call a "chamber". After one quarter of a cycle we move to diagram B and find that chamber 1 has rotated clockwise and decreased in size; this continues through diagrams C and D. Then, after one complete cycle we are back at diagram A, but the gas that started life in chamber 1 is now in chamber 2. As we continue through the second cycle, chamber 2 is compressed further and finally, at the beginning of the third cycle, it turns into chamber 3 which now

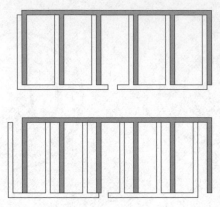

Fig. 1.2. Side view of a scroll compressor

opens up to a vent at the centre of the scroll through which the compressed gas escapes.

Today, scroll compressors are widely used for compressing refrigerants in air conditioners and industrial refrigerators, and somewhat less widely for air compression. There are many advantages over the traditional pump-type design. For example, only a small number of moving parts and no valves are required, and the rotary motion can be completely balanced, reducing vibration and noise. The biggest problem with them is leakage of gas, resulting in reduced efficiency. Thus, and when possible new designs are considered, controlling the leakage is of paramount importance.

Another problem with the traditional design is that the compression takes place rather slowly, so that a large number of turns is required to achieve the high compression ratios demanded by customers. Unfortunately, for a given total cross-sectional area, increasing the tightness of the spiral (i) decreases the "choke volume" of gas that can be ingested in each cycle (i.e. the volume of chamber 1 in figure 1.1Λ), (ii) makes the job of machining the scrolls more difficult, and (iii) increases the rate of leakage. A possible solution to this problem is to use a spiral different from the circle involute shown in figure 1.1, specifically one that gives more rapid compression without reducing the intake volume.

The questions asked of the Study Group by DANFOSS were:

- What different spiral geometries can be used to make a workable scroll compressor?
- How do different scroll geometries affect the performance and efficiency of the compressor?

- Can we devise a "cost-function" that can be used to optimise over different proposed designs?

The problem divides naturally into two lines of attack. First, it is an exercise in the geometry of plane curves to determine the important characteristics (choke volume, compression ratio, etc.) of a compressor made from a given spiral. Second, once these geometrical properties have been determined, the performance and efficiency must be found by considering the fluid mechanics of the gas. It is the second aspect of the problem (i.e. the final two questions posed above) that is the subject of this case study. We will suppose that the chamber volumes and all other important geometrical parameters are known, and attempt to assess their effects on the behaviour of the gas in the compressor. For more details about the geometrical aspects of the problem, see the study group report [2]. (It was found that the important properties of a compressor design are conveniently found by expressing the spirals in the "natural" arc-length–angle formulation.)

The first consideration is the leakage between adjoining chambers, which forms the subject of section 1.2. Then in section 1.3 we derive equations for the temperature and pressure of the gas in each chamber. The leakage found in section 1.2 acts as a coupling mechanism between a chamber and its neighbours, and a fully coupled model is described in section 1.4. Some numerical results for a simple realisation of this model are presented in section 1.5, and the conclusions are drawn in section 1.6, where some further projects for the interested reader are also suggested.

1.2 Leakage between Chambers

Ideally, the two halves of a scroll compressor remain perfectly in contact as they rotate. In reality it is not practical to machine them accurately enough for this to be the case, and instead there is a narrow gap. Typically the gap is around one micron across; this may be increased by wear and/or poor machining, and it is known that, if it reaches around eight microns, the compressor becomes useless. In this section we analyse the flow through this gap in an attempt to determine the leakage between neighbouring chambers in the compressor. A schematic diagram of the geometry is given in figure 1.3. Here a gap of typical thickness d and length L separates two adjacent chambers containing gas at pressures P_1 and P_2. The length scale L appears to be somewhat arbitrary; a typical value of 1 cm was obtained by "eyeballing" a sample compressor (whose total diameter was of the order of 20 cm) supplied by DANFOSS.

A local coordinate system is adopted with x pointing along the gap and y pointing across it. With respect to these coordinates we denote the bottom

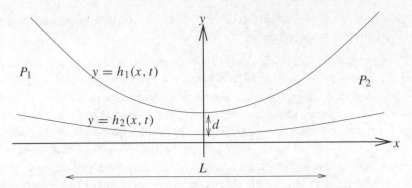

Fig. 1.3. Schematic diagram of the gap between two chambers

and top scroll walls (both of which are moving as the compressor rotates) by $y = h_1(x, t)$ and $y = h_2(x, t)$.

1.2.1 Governing Equations and Boundary Conditions

The governing equations for the flow are the compressible Navier–Stokes equations (see [4])

$$\rho_t + \operatorname{div}(\rho \boldsymbol{u}) = 0, \tag{1.1}$$

$$\rho \frac{D\boldsymbol{u}}{Dt} = -\operatorname{grad} p + (\lambda + \mu)\operatorname{grad} \operatorname{div} \boldsymbol{u} + \mu \nabla^2 \boldsymbol{u}, \tag{1.2}$$

where ρ, $\boldsymbol{u}$, and p are the density, velocity, and pressure fields, λ and μ are the dilatational and shear viscosities, a D/Dt is the usual convective derivative. Equation (1.1) represents conservation of mass, and must be satisfied by *any* continuum with velocity $\boldsymbol{u}$ and density ρ, while (1.2) is the generalisation of the usual incompressible Navier–Stokes equation, which is recovered if div $\boldsymbol{u}$ is set to zero.

In contrast with the incompressible case, (1.1, 1.2) is *not* a closed system since the unknowns ρ, $\boldsymbol{u}$, and p outnumber the equations by one. Thus it is necessary to consider also the *energy equation* (see [4])

$$\rho c_v \frac{DT}{Dt} + p \operatorname{div} \boldsymbol{u} = k \nabla^2 T + \Phi, \tag{1.3}$$

where T is absolute temperature, k and c_v are the thermal diffusivity and specific heat at constant volume of the gas, and Φ is the *dissipation*,

$$\Phi = \lambda (\operatorname{div} \boldsymbol{u})^2 + \frac{\mu}{2} \sum_{i,j=1,2} \left(\frac{\partial u_i}{\partial x_j} + \frac{\partial u_j}{\partial x_i} \right)^2. \tag{1.4}$$

The terms in (1.3) represent respectively (i) rate of change of thermal energy, (ii) work done by pressure, (iii) diffusion of heat, and (iv) heat generated by viscous effects.

Finally, the system is closed by specifying an equation of state. Treating the air as a perfect gas, we set

$$p = \rho R T, \tag{1.5}$$

where the so-called gas constant R can be written as

$$R = c_p - c_v,$$

and c_p is the specific heat at constant pressure. It is convenient to substitute (1.1) and (1.5) into the left-hand side of (1.3) to obtain the energy equation in the form

$$\frac{1}{\gamma - 1} \frac{Dp}{Dt} - \frac{\gamma}{\gamma - 1} \frac{p}{\rho} \frac{D\rho}{Dt} = k\nabla^2 T + \Phi, \tag{1.6}$$

where γ is the ratio of specific heats,

$$\gamma = \frac{c_p}{c_v},$$

which takes a value very close to 1.4 for air.

As an aside we note that, in many gas dynamics problems, the right-hand side of (1.6) is negligible compared with the left-hand side. If so, (1.6) implies that $D/Dt(p/\rho^\gamma) = 0$, and often one can deduce further that p is given as a function of density by $p = k\rho^\gamma$ for some constant k; such flows are described as *homentropic*. This approach does not work for the current problem, since we will find that the dissipation term on the right-hand side of (1.6) is necessarily the same order as the left-hand side.

On the upper and lower walls we specify the velocity $\boldsymbol{u} = (u, v)$ of the gas to be the same as that of the scroll:

$$u = u_i, \quad v = \frac{\partial h_i}{\partial t} + u_i \frac{\partial h_i}{\partial x}, \quad \text{on } y = h_i(x, t); \quad i = 1, 2. \tag{1.7}$$

Here the horizontal velocity u_i and position h_i of each wall are determined by the prescribed shape and motion of the scroll. Note that (1.7) holds for *any* such motion, even if the walls were flexible (e.g. peristaltic action), although only rigid walls are relevant to our application.

We now integrate (1.1) with respect to y and apply the boundary conditions (1.7) to obtain an integrated equation representing total conservation of mass:

$$m_t + q_x = 0, \quad \text{where } m = \int_{h_1}^{h_2} \rho \, dy, \quad q = \int_{h_1}^{h_2} \rho u \, dy. \tag{1.8}$$

The quantities m and q represent respectively the mass density of gas in, and flux of gas through, each cross section.

As boundary conditions for the temperature, we assume that the scroll walls are thermally insulated, so that the normal derivative of T is zero, that is,

$$\frac{\partial T}{\partial y} = \frac{\partial h_i}{\partial x} \frac{\partial T}{\partial x} \text{ on } y = h_i(x, t); \quad i = 1, 2. \tag{1.9}$$

We also have to match the solution in the gap with the prescribed pressure in the chamber on either side:

$$p \to P_1 \text{ as } x \to -\infty, \quad p \to P_2 \text{ as } x \to +\infty. \tag{1.10}$$

1.2.2 Dimensionless Parameters

Now we compare the sizes of the different terms in equations (1.1–1.6) to see what simplifications can be made. Typical values for the physical parameters in the problem are given in table 1.1. First we note that the geometry is very slender: the gap is typically much longer than it is thick. This feature is characterised by the *slenderness parameter*

$$\epsilon = \frac{d}{L} \approx 10^{-4}, \tag{1.11}$$

and the fact that this is very small will enable us to use *lubrication theory*: a great simplification of the Navier–Stokes equations.

Now, given the pressure drop ΔP across the gap, we can deduce a typical gas velocity U by balancing the pressure gradient with viscous drag in (1.2):

$$U = \frac{d^2 \Delta P}{\mu L}.$$

Table 1.1. *Estimated Parameter Values for Leakage Between Chambers in a Scroll Compressor*

Property	Symbol	Approx. Value	Units
Viscosity	μ	10^{-5}	Pa s
Density	$\bar{\rho}$	1	Kg m^{-3}
Thermal diffusivity	k	2.5×10^{-2}	$\text{J m}^{-1}\text{s}^{-1}\text{K}^{-1}$
Specific heat	c_v	10^3	$\text{J Kg}^{-1}\text{K}^{-1}$
Gap thickness	d	10^{-6}	m
Contact length	L	10^{-2}	m
Rotation frequency	ω	50	s^{-1}
Pressure drop	ΔP	10^6	Pa

Another dimensionless parameter is the ratio between this quantity and the velocity due to rotation of the compressor:

$$\Omega = \frac{\omega L}{U} = \frac{\mu \omega L^2}{d^2 \Delta P} \approx 5 \times 10^{-2}. \tag{1.12}$$

In general the gas flow can be decomposed into (i) a "Couette flow" caused by relative tangential motion of the walls, (ii) a "squeeze film" due to movement of the walls towards or away from each other, and (iii) a "Poiseuille flow" in which gas is forced through the gap by the imposed pressure difference. If, as appears to be the case, Ω is small, this says that (iii) dominates over (i) and (ii).

Now, using U as the velocity scale we compare the left- and right-hand sides of (1.2) to determine the importance of inertia in the problem. As usual in lubrication-type problems, the corresponding dimensionless parameter is the *reduced Reynolds number*, which is the classical Reynolds number reduced by a factor of ϵ^2:

$$\text{Re}^* = \frac{\bar{\rho} U L \epsilon^2}{\mu} = \frac{\bar{\rho} d^4 \Delta P}{\mu^2 L^2} \approx 10^{-4}. \tag{1.13}$$

Since Re^* is small, viscous effects dominate inertia; that is, the left-hand side of (1.2) can safely be neglected.†

Similarly, we compare the left- and right-hand sides of (1.6) using the *reduced Peclet number*:

$$\text{Pe}^* = \frac{\bar{\rho} c_v U L \epsilon^2}{k} = \frac{\bar{\rho} c_v d^4 \Delta P}{\mu k L^2} \approx 10^{-4}.$$

It is no accident that $\text{Pe}^* \approx \text{Re}^*$ since their ratio, the Prandtl number $\text{Pr} = k/(\mu c_v)$, is close to unity for air.

We can immediately deduce a typical rate of leakage from the velocity scale U. The rate at which gas is lost through the channel is of order Ud, and so the cumulative loss over a cycle is typically Ud/ω. We simply have to compare this with the original area of a chamber to obtain the relative loss of gas due to leakage:

$$\text{relative loss} = \frac{d^3 \Delta P}{\mu \omega L^3} = \frac{\epsilon}{\Omega}. \tag{1.14}$$

---

† We can also interpret Re^* as the square of a typical Mach number:

$$\text{Re}^* = \frac{U^2}{c^2}, \quad \text{where } c^2 = \frac{\Delta P}{\bar{\rho}},$$

and hence deduce that the flow is wholly subsonic.

The values given in table 1.1 suggest that this is rather small: about 0.2%. However, it is clearly highly sensitive to increases in d, and if we set $d = 8 \, \mu$m, then the typical relative loss is dramatically increased to around 100%. This is in encouraging agreement with the experimental observations noted earlier.

1.2.3 Dimensionless Equations

In nondimensionalising the equations (1.1–1.6) we utilise the slenderness of the geometry and the difference between the velocity scales for the gas and for the compressor. Thus we set

$$x = L x', \quad y = d y', \quad u = U u', \quad v = \epsilon U v', \quad t = t'/\omega,$$
$$h_i = d h_i', \quad u_i = \omega L u_i', \quad p = \Delta P p', \quad \rho = \bar{\rho} \rho', \quad T = \Delta P T'/(\bar{\rho} c_v).$$
$$(1.15)$$

Henceforth we drop the primes and proceed with the dimensionless variables. The Navier–Stokes equations (1.2), up to order Re* and ϵ^2, reduce to the *lubrication equations* (see [4])

$$p_y = 0, \quad u_{yy} = p_x.$$
$$(1.16)$$

Thus,

$$u = \frac{p_x}{2} y^2 + A y + B,$$
$$(1.17)$$

for some $A(x, t)$ and $B(x, t)$, found using the dimensionless version of (1.7) to be given by

$$A = \frac{\Omega (u_2 - u_1)}{h} - \frac{p_x (h_1 + h_2)}{2}, \quad B = \frac{\Omega (h_2 u_1 - h_1 u_2)}{h} + \frac{p_x h_1 h_2}{2},$$
$$(1.18)$$

where h is the gap thickness:

$$h = h_2 - h_1.$$

Now consider the dimensionless version of the energy equation (1.6) and thermal boundary condition (1.9), taking only the terms at leading order in ϵ:

$$\frac{1}{\gamma - 1} \left(\Omega p_t + u p_x + v p_y \right) - \frac{\gamma}{\gamma - 1} \frac{p}{\rho} \left(\Omega \rho_t + u \rho_x + v \rho_y \right) - u_y^2 + O(\epsilon^2)$$

$$= \frac{1}{\text{Pe}^*} \left(T_{yy} + O(\epsilon^2) \right),$$
$$(1.19)$$

$$T_y = O(\epsilon^2) \quad \text{on } y = h_1, h_2.$$
$$(1.20)$$

Now we use the fact that Pe* is much smaller than one to expand T as an asymptotic expansion in powers of Pe*:

$$T = T_0 + \text{Pe}^* T_1 + \cdots .$$

Notice that, since $\epsilon^2 \ll \text{Pe}^*$, it makes sense to keep terms of order Pe* while neglecting those of order ϵ^2.

To lowest order in Pe*, (1.19, 1.20) reduces to the trivial homogeneous Neumann problem

$$\frac{\partial^2 T_0}{\partial y^2} = 0, \quad \frac{\partial T_0}{\partial y} = 0 \quad \text{on} \quad y = h_1, h_2, \tag{1.21}$$

from which we can deduce only that T_0 is independent of y. But from (1.16) we know that p is also independent of y, and thus (1.5) implies that ρ must likewise be a function only of x and t to lowest order.

We can obtain no more information about the functions $p(x, t)$ and $\rho(x, t)$ from the leading-order problem (1.21). This situation, where the leading-order problem admits nonunique solutions, arises quite often. The way to resolve the nonuniqueness is to proceed to higher order in Pe* and consider the problem for T_1, namely†

$$\frac{\partial^2 T_1}{\partial y^2} = \frac{1}{\gamma - 1} (\Omega p_t + u p_x) - \frac{\gamma}{\gamma - 1} \frac{p}{\rho} (\Omega \rho_t + u \rho_x) - u_y^2, \tag{1.22}$$

$$\frac{\partial T_1}{\partial y} = 0 \quad \text{on} \quad y = h_1, h_2. \tag{1.23}$$

Now we have an inhomogeneous Neumann problem, which can only admit solutions for T_1 if a *solvability condition* is satisfied. In this case, by integrating (1.22) with respect to y and applying (1.23), we obtain the required relation between $p(x, t)$ and $\rho(x, t)$, namely

$$\Omega p_t + \bar{u} p_x - \frac{\gamma p}{\rho} (\Omega \rho_t + \bar{u} \rho_x) = (\gamma - 1)\overline{u_y^2}, \tag{1.24}$$

where $\bar{}$ denotes the cross-sectional average:

$$\bar{\cdot} = \frac{1}{h} \int_{h_1}^{h_2} \cdot \, dy.$$

Now we simply substitute in the analytic form (1.17) of u to obtain

$$\Omega \left(p_t - \frac{\gamma p}{\rho} \rho_t \right) = \frac{\gamma h^2 p_x}{12} \left(p_x - \frac{p}{\rho} \rho_x \right). \tag{1.25}$$

---

† Strictly speaking we should expand p, ρ, and u in powers of Pe* also, but for ease of presentation we do not bother; in effect we use p, ρ, and u as shorthand for p_0, ρ_0, and u_0.

A second equation linking p and ρ is obtained by substituting (1.17) into the dimensionless form of (1.8), the result of which is

$$\Omega(\rho h)_t + q_x = 0, \quad \text{where } q = \frac{\Omega \rho h(u_1 + u_2)}{2} - \frac{\rho p_x h^3}{12}. \tag{1.26}$$

1.2.4 Solution in the Quasi-Steady Limit

Equations (1.25, 1.26) form a closed leading-order system for ρ and p; recall that u_1, u_2 and h are prescribed functions of x and t. They can be simplified further by taking the *quasi-steady limit* $\Omega \to 0$. Then, from (1.25) we deduce that

$$\frac{\partial}{\partial x} \left(\frac{p}{\rho} \right) = 0, \tag{1.27}$$

which implies that the gas is *isothermal*. So in (1.26) we can set $\rho = p/(T(\gamma - 1))$, where T is independent of x. Then, by setting $\Omega = 0$ in (1.26) we find that the flux

$$q = -\frac{h^3 p p_x}{12T(\gamma - 1)} \tag{1.28}$$

is a function only of t. Now we simply divide by h^3 and integrate with respect to x from $-\infty$ to $+\infty$, applying the matching conditions (1.10), to obtain

$$q = \frac{P_1^2 - P_2^2}{24T(\gamma - 1)} \left(\int_{-\infty}^{\infty} \frac{dx}{h^3} \right)^{-1}. \tag{1.29}$$

This tells us how the flux, i.e. the leakage through the gap, depends on the pressure in the chamber on either side and the geometry of the channel.

It is worth emphasising the difference between this *quasi-steady* solution and a steady-state solution in which all the dependent variables are assumed to be independent of t. We have taken a limit in which the time derivatives in (1.25, 1.26) can be neglected, but p and ρ may still depend on t. In particular, in (1.29) P_1, P_2, T, and h can all be expected to be functions of time.†

Since the minimum gap thickness is very small, the integral in (1.29) is dominated by the behaviour of h near its minimum. In a neighbourhood of this point, we can approximate h by a quadratic function, say

$$h = d(t) + \frac{\kappa(t)x^2}{2},$$

---

† The quasi-steady limit reflects the fact that these variables vary only slowly with t: the parameter Ω can be thought of as the ratio of the time scale (L/U) associated with convection in the gas flow to the time scale $(1/\omega)$ over which P_1, P_2, T, etc. vary.

where κ is the *difference between the curvatures* of the two channel walls at their closest point. Then

$$\int_{-\infty}^{\infty} \frac{dx}{h^3} = \frac{3\pi}{4d^{5/2}\sqrt{2\kappa}}.$$

1.3 Conservation Equations for the Chambers

Now we consider the conservation of mass and energy for a single chamber of gas. Since the Reynolds number on the scale of a chamber is large, it is usual to assume that the gas in each chamber is turbulent and thus well mixed. Hence we can associate a spatially uniform temperature $T(t)$ and pressure $P(t)$ with the mass $M(t)$ of gas in the chamber, which has a given volume $V(t)$.

First, note that the ratio of kinetic to thermal energy is of order

$$\frac{\text{kinetic energy}}{\text{thermal energy}} \sim \frac{\bar{\rho}\omega^2 L^2}{\Delta P} = \Omega^2 \text{Re}^* \ll 1.$$

Therefore we neglect kinetic energy throughout, so that the internal energy in the chamber is simply

$$E = Mc_v T. \tag{1.30}$$

The internal energy is changed due to (see e.g. [5]):

- The work done by changes in the volume V of the chamber which, recall, is assumed to be a prescribed function of time. The work done is given by $-P\,dV/dt$.
- The energy transported into and out of the chamber through the gaps on either side. Assuming the flow through these gaps to be adiabatic, the energy transported by a mass flux q is qH, where H is the *enthalpy*, equal to $c_p T$.
- Dissipation and energy losses to the external environment, which we denote by $\dot{Q}$.

Putting all these together, we obtain

$$\frac{dE}{dt} = -P\frac{dV}{dt} + q_i c_p T_i - q_o c_p T + \dot{Q}, \tag{1.31}$$

where the subscripts i and o correspond to flow into and out of the chamber respectively. Notice that the temperature of gas flowing out of the chamber is the same as that of the gas in the chamber, T.

Henceforth we neglect $\dot{Q}$, although thermal interaction with the surroundings might be included in a more refined model. Therefore, substituting

(1.30) into (1.31) and rearranging, we obtain an ordinary differential equation relating P and T:

$$V\frac{dP}{dt} = \gamma R(q_i T_i - q_o T) - \gamma P \frac{dV}{dt}. \tag{1.32}$$

Next we consider conservation of mass for the chamber. This simply states that the mass change is equal to the flux into the chamber minus that out:

$$\frac{dM}{dt} = q_i - q_o. \tag{1.33}$$

But $M = \rho V$, where the density ρ is given in terms of P and T by using the equation of state (1.5). Thus (1.33) can be rearranged to a second equation relating P and T:

$$\frac{dT}{dt} = \frac{T}{P}\frac{dP}{dt} + \frac{T}{V}\frac{dV}{dt} - \frac{RT^2}{PV}(q_i - q_o). \tag{1.34}$$

Now the idea is as follows. Given the pressure P and temperature T in two neighbouring chambers, and the geometry of the gap between them, we can evaluate the flux from one to the other using (1.29). Then for each chamber we have the two differential equations (1.32, 1.34) for P and T, which are coupled to the corresponding equations for the neighbouring chambers by the fluxes q on either side.

1.4 The Coupled Problem

The situation is depicted schematically in figure 1.4. We have a series of chambers, in the nth of which the gas is at pressure P_n and temperature T_n. The flux between the nth and $(n+1)^{\text{th}}$ chamber is denoted by q_n, with the sign convention that $q_n > 0$ if gas flows *from* chamber n to chamber $(n+1)$.

From the theory of section 1.2, we know that in each of the gaps between the chambers the temperature is constant, but the value of that constant is

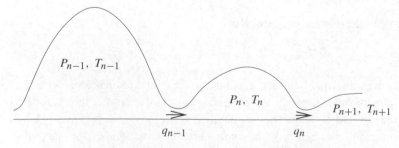

Fig. 1.4. Schematic diagram of the coupling between neighbouring chambers

determined by the sign of q: if $q_n > 0$ then the gas transported by q_n has temperature T_n, while if $q_n < 0$ it is the following expression for q_n:

$$q_n = \frac{\left(P_n^2 - P_{n+1}^2\right) d^{5/2} \sqrt{\kappa}}{9\pi \sqrt{2}\, \mu R} \begin{cases} 1/T_n & \text{if } P_n > P_{n+1}, \\ 1/T_{n+1} & \text{if } P_n < P_{n+1}. \end{cases} \tag{1.35}$$

Similarly, when we write down the conservation equations (1.32, 1.34) for the nth chamber, whether each of q_{n-1} and q_n qualifies as flux *into* (q_i) or *out of* (q_o) the chamber depends on its sign. The resulting equations can conveniently be written in the form

$$\frac{\dot{P}_n}{P_n} = \frac{\gamma R}{P_n V_n} \left\{ q_{n-1} T_n - q_n T_{n+1} + q_{n-1}^+ (T_{n-1} - T_n) \right.$$

$$\left. + q_n^+ (T_{n+1} - T_n) \right\} - \frac{\gamma \dot{V}_n}{V_n}, \tag{1.36}$$

$$\frac{\dot{T}_n}{T_n} = -\frac{(\gamma - 1)\dot{V}_n}{V_n} + \frac{R}{P_n V_n} \left\{ q_n T_n - \gamma q_n T_{n+1} \right.$$

$$\left. + (\gamma - 1) q_{n-1} T_n + \gamma q_{n-1}^+ (T_{n-1} - T_n) + \gamma q_n^+ (T_{n+1} - T_n) \right\}, \tag{1.37}$$

where q^+ denotes the positive part of q:

$$q^+ = \begin{cases} q & \text{if } q > 0, \\ 0 & \text{if } q < 0. \end{cases}$$

To solve the dynamical system (1.36, 1.37) we need to apply "end conditions" at the outermost chambers. Suppose there are N chambers altogether, with the first and N^{th} open to reservoirs at *given* pressure and temperature P_0, P_N and T_0, T_N. (Note that T_0 and/or T_N need only be specified if $P_0 > P_1$ and/or $P_N > P_{N-1}$, so that gas flows *into* the adjoining chambers.) We also have to specify P_n and T_n for $n = 1, \dots, N - 1$ at $t = 0$. Finally, there is a complicated closure condition associated with the periodicity of the motion. Roughly speaking, it is clear that after a complete cycle, what was the nth chamber has now become the $(n + 1)^{\text{th}}$ chamber. The way in which this condition is implemented in practice should be made clear by the following outlined solution procedure.

(i) Suppose that P_n and T_n are given at $t = 0$ for $n = 1, \dots, N$.

(ii) Integrate the coupled ordinary differential equations (1.36, 1.37) forward through one complete cycle, using the specified values of P_0, T_0, P_N, T_N.

(iii) Set

$$\{V_n, P_n, T_n\}_{\text{new}} = \{V_{n-1}, P_{n-1}, T_{n-1}\}_{\text{old}}.$$

(iv) Go to step (ii).

The desired final result is a periodic solution, in which the "new" and "old" values in step (iii) above are identical, that is

periodic solution $\Rightarrow$

$$\{V_n(t), P_n(t), T_n(t)\} \equiv \{V_{n-1}(t+\tau), P_{n-1}(t+\tau), T_{n-1}(t+\tau)\},$$

where $\tau = 2\pi/\omega$ is the period of the motion. However, when there is strong coupling between the chambers, it is far from clear that this periodic solution is unique or stable. Indeed, for a high-dimensional nonlinear dynamical system such as this, we might expect to see rather complicated dynamics in general.

1.4.1 The Small Coupling Limit

If the leakage is relatively small (which it should be for any worthwhile compressor), we can perturb about the zero-leakage solution, obtained by solving (1.36, 1.37) with $q_n \equiv 0$:[†]

$$P_n = P_0 \left(\frac{V_0}{V_n}\right)^{\gamma}, \quad T_n = T_0 \left(\frac{V_0}{V_n}\right)^{\gamma-1}. \tag{1.38}$$

Here V_0 is the *choke volume*: the volume of gas ingested by the compressor at the outset of the cycle (the volume marked "1" in figure 1.1A). Then the lowest-order fluxes are obtained by substituting (1.38) into the expression (1.35) for q_n.

Interestingly, if the compressor is in a periodic state, we only need to find q_0 to evaluate the total leakage: in its first cycle, a chamber *gains* $-q_1$ and loses $-q_0$. Then, in the next cycle, it *loses* $-q_1$ and gains $-q_2$. Over the lifetime of a chamber, all the intermediate fluxes cancel each other out, so only q_0 remains. In the small-q limit, this is readily evaluated:

$$-q_0 = \frac{P_0^2 d^{5/2}\sqrt{\kappa}}{9\pi\sqrt{2}\,\mu R T_0} \left(\frac{V_0}{V_1}\right) \left[\left(\frac{V_0}{V_1}\right)^{\gamma} - \left(\frac{V_1}{V_0}\right)^{\gamma}\right]. \tag{1.39}$$

[†] When there is no leakage, the *mass* and *entropy* of the gas in each chamber are preserved: see e.g. [5].

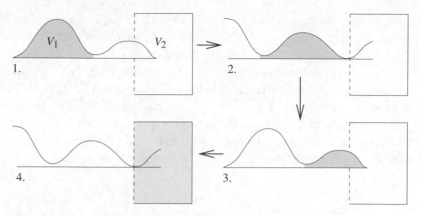

Fig. 1.5. Schematic diagram of the compression of a single chamber (of volume V_1) which opens up into a reservoir of volume V_2

Thus, assuming that d and the other parameters in (1.39) are constant (and in any case beyond our control), we obtain a functional form of the total leakage:

$$\text{total leakage} \propto l := \int_0^\tau \sqrt{\kappa} \left(\frac{V_0}{V_1}\right) \left[\left(\frac{V_0}{V_1}\right)^\gamma - \left(\frac{V_1}{V_0}\right)^\gamma\right] dt. \qquad (1.40)$$

We can use this functional as part of a cost function in comparing proposed new compressor designs, since all the variables on the right hand side of (1.40) can readily be evaluated for any given scroll geometry. Without performing any detailed calculations, we can immediately deduce some desirable design properties that will reduce leakage:

- The contact should be as *flat* as possible, i.e. κ should be minimised
- The volume should be *reduced gradually on the first cycle*. This follows from the observation that only q_0 is relevant to the total mass loss.

1.5 Numerical Results

In this section we present some preliminary numerical simulations of the system (1.36, 1.37). We consider the simple configuration shown schematically in figure 1.5. Here, at the beginning of the cycle (diagram 1) the shaded volume of gas V_1 is taken in from the atmosphere. As V_1 decreases (diagram 2) the gas is compressed until the cycle is completed (diagram 3). Then the gas is released into a reservoir of *constant* volume V_2 (diagram 4). The idea is that we run the simulation through several such cycles and see what pressure we can achieve in the reservoir; this seems like a good measure of the efficacy of the compressor.

Equations (1.36, 1.37) with q_0 and q_1 given by (1.35), $q_2 \equiv 0$ and $n = 1, 2$ provide a closed system for P_1, P_2, T_1, and T_2, given the inlet pressure P_0 and temperature T_0. We initiate the calculations with $P_1 = P_2 = P_0$, $T_1 = T_2 = T_0$. As outlined in section 1.4, at the end of each cycle we perform a replacement algorithm corresponding to (i) a new chamber of atmospheric gas forming in the "new" V_1; (ii) the "old" V_1 discharging into V_2. For the latter we find the new values of P_2 and T_2 by setting the mass and energy in V_2 after the discharge equal to the total mass and energy in V_1 and V_2 immediately prior to the discharge. Thus, after the nth cycle, we set

$$P_1(n\tau+) = P_0, \qquad P_2(n\tau+) = \left[\frac{P_1 V_1 + P_2 V_2}{V_1 + V_2} \right] (n\tau-),$$

$$T_1(n\tau+) = T_0, \qquad T_2(n\tau+) = \left[\frac{(P_1 V_1 + P_2 V_2)T_1 T_2}{P_1 V_1 T_1 + P_2 V_2 T_2} \right] (n\tau-). \tag{1.41}$$

In all the calculations to follow the parameter values are set as follows,

$$\mu = 1, \quad R = 1, \quad \gamma = 1.4, \quad \kappa = 1,$$
$$P_0 = 1, \quad T_0 = 1, \quad V_2 = 10, \quad \tau = 1,$$

and we examine the effects of varying the "leakage parameter" d and the form of $V_1(t)$. We start with the simplest case in which V_1 is linear in t, say

$$V_1 = 1 - \alpha t,$$

where $\alpha \in (0, 1)$; $\alpha = 0$ implies a compression ratio of one, while as $\alpha \to 1$ the compression ratio goes to infinity. (In [2] it is shown that this linear variation of V_1 with time corresponds to the traditional circle involute design shown in figure 1.1.)

In figure 1.6 we plot the pressure in the chamber and the reservoir versus time for the case $\alpha = 0.5$, $d = 1$. We can see how P_1 increases during each cycle and is reset each time a new cycle begins. In the reservoir, P_2 varies only slightly during any cycle, and is incremented gradually at each discharge. Closer examination reveals that, because of leakage between the chamber and the reservoir, P_2 *decreases* when $P_2 > P_1$ and *increases* when $P_2 < P_1$. For these parameter values, the system appears to settle down to a periodic state after around 60 cycles, with the pressure in the reservoir enhanced by a factor of just over 2. From the calculations, we can also obtain the temperature variations, but we do not bother to show these, since they are irrelevant to the total energy stored in the reservoir, which is proportional to $P_2 V_2$.

In figure 1.7 we present the corresponding results when $\alpha = 0.8$, so the compression ratio is five, compared with two in the previous calculation. As expected, the increased compression ratio leads to a higher final pressure,

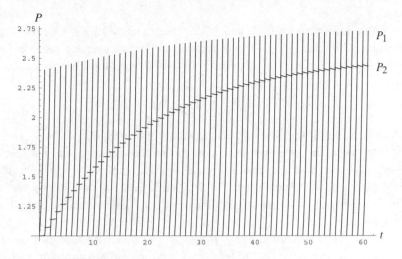

Fig. 1.6. Pressures P_1 (in the compression chamber) and P_2 (in the pressurised reservoir) versus time. The volumes are $V_1 = 1 - 0.5\,t$ and $V_2 = 10$, and the leakage parameter is $d = 1$

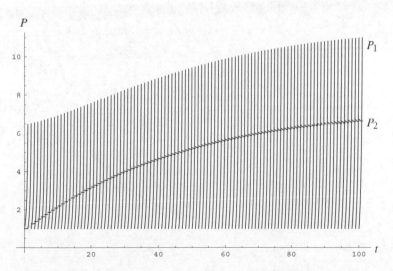

Fig. 1.7. Pressures P_1 (in the compression chamber) and P_2 (in the pressurised reservoir) versus time. The volumes are $V_1 = 1 - 0.8\,t$ and $V_2 = 10$, and the leakage parameter is $d = 1$

although the system also takes somewhat longer to converge to its periodic state.

In the light of figures 1.6 and 1.7, it is of interest to ask how the final pressure achieved in the reservoir depends on the compression ratio, i.e. on α. As a measure of this we take the average of P_2 over the 100th cycle:

$$\bar{P}(100) = \int_{100}^{101} P_2 \, dt,$$

and plot the result versus α for various values of d in figure 1.8. Not surprisingly the pressure achieved increases as α is increased and as d is decreased. However, for sufficiently small d, decreasing d still further does not appear to have much effect; the graphs for $d = 0.25$ and $d = 0.5$ are indistinguishable. This is explained by examination of the transients, which makes it clear that, for these small values of d, P_2 has yet to equilibrate after 100 cycles.

So, we have shown the rather obvious results that the effectiveness of a compressor can be enhanced by increasing the compression ratio and by reducing the leakage, although these also make the convergence to maximum compression slower. Next we would like to compare compressors with the same compression ratio but different histories $V_1(t)$. Therefore we consider the family

$$V_1 = 1 - \beta t + (V_f + \beta - 1)t^2,$$

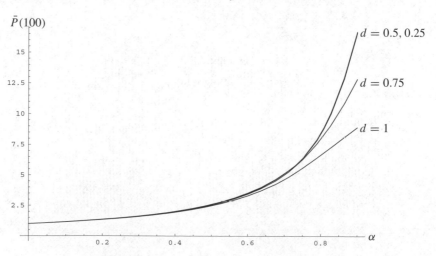

Fig. 1.8. Average reservoir pressure over the 100th cycle versus compression parameter α. The volumes are $V_1 = 1 - \alpha t$ and $V_2 = 10$, and the leakage parameter d takes the values 0.25, 0.5, 0.75, 1. Notice that the curves for $d = 0.25$ and $d = 0.5$ are indistinguishable

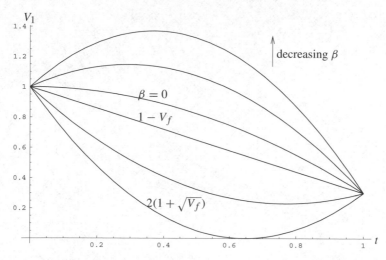

Fig. 1.9. The family of curves $V_1 = 1 - \beta t + (V_f + \beta - 1)t^2$ for various values of β (here $V_f = 0.3$). The value $\beta = 1 - V_f$ gives a straight line joining $V_1 = 1$ at $t = 0$ to $V_1 = V_f$ at $t = 1$; $\beta = 0$ has zero gradient at $t = 0$; $\beta = 2(1 + \sqrt{V_f})$ is the value at which V_1 first reaches zero for $t \in (0, 1)$

shown in figure 1.9, where V_f is the final volume (so the compression ratio is $1/V_f$) and β changes the volume history for a fixed V_f; $\beta = 1 - V_f$ gives the linear $V_1(t)$, i.e. constant compression rate, considered previously. Broadly speaking, if β is decreased, the compression is slower initially and accelerates towards the end of the cycle, and vice versa.

First we check how our theoretical approximate leakage l given in (1.40) varies with β. In figure 1.10 we plot l versus β for different values of V_f. We observe that the theoretical leakage for a given volume ratio is reduced if the compression is slow at the beginning and faster at the end. This echoes the suggestion at the end of section 1.4 that the volume should be reduced gradually at the start. There is a surprise, however, in that l can be *negative* if β is sufficiently large and negative. This corresponds to large positive excursions in V_1 (see figure 1.9), as a result of which the compressor acts like a bellows, sucking extra gas into the chamber.

This conclusion is backed up by figure 1.11 in which we plot $P(100)$ from our simulation versus β for different values of V_f (and with $d = 1$). Here the behaviour in general is extremely interesting: depending on the compression ratio, $\bar{P}$ may be an increasing or decreasing function of β, or may vary nonmonotonically. However, for the large compression ratios which are likely to be of most practical interest, the optimal performance (i.e. largest possible value of $\bar{P}(100)$) is obtained by making β as large and negative as possible.

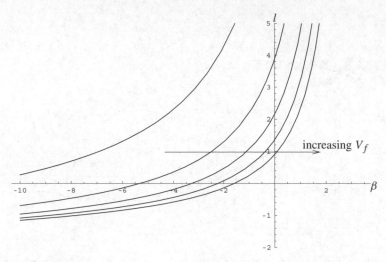

Fig. 1.10. Theoretically predicted leakage $l = \int_0^1 (V_1^{-\gamma-1} - V_1^{\gamma-1})\, dt$ for the family of volume histories depicted in figure 1.9; $V_1 = 1 - \beta t + (V_f + \beta - 1)t^2$ with $V_f = 0.1$, $0.2, 0.3, 0.4, 0.5$

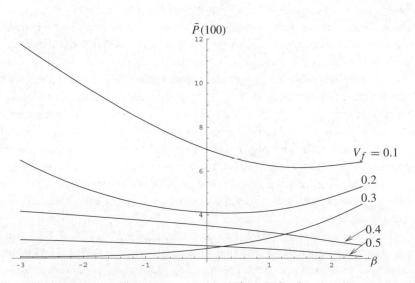

Fig. 1.11. Average reservoir pressure over the 100th cycle for $d = 1$ and the family of volume histories depicted in figure 1.9; $V_1 = 1 - \beta t + (V_f + \beta - 1)t^2$ with $V_f = 0.1$, $0.2, 0.3, 0.4, 0.5$

1.6 Conclusions

Here we summarise the work presented in this case study and suggest some open questions that the interested reader might consider further.

We have used compressible lubrication theory to obtain a theoretical prediction of the leakage between adjoining chambers in a scroll compressor. One surprising outcome of the analysis is that the gas flowing through the narrow gap between one chamber and the next is *isothermal*. In traditional inviscid-gas dynamics one would expect the temperature to decrease as the gas accelerates through the gap, while classical lubrication theory would predict an increase in temperature due to viscous dissipation. Remarkably these two effects appear to cancel each other out exactly.

- Is there a simple physical explanation for this result?

Our result for interchamber leakage was then incorporated in a coupled model for the pressure and temperature of the gas in each chamber. The model takes the form of a dynamical system.

- What can be said about the general properties of the dynamical system (1.36, 1.37)? For example, is the desired periodic solution linearly stable?

Our analysis was limited to examination of the case in which the leakage is small. In this limit we obtained an approximate measure of the total losses due to leakage, as a functional of the geometry of the compressor. This could in future be included in a cost function for evaluating proposed new compressor designs.

- What other parameters should be included in the cost function? How might an optimisation procedure be implemented?

We performed some numerical simulations of a simple compressor with just one chamber pumping gas into a sealed reservoir. From these we were able to test the effects on compressor performance of varying the leakage rate, the volume ratio, and the volume history. The most intriguing possibility suggested by the simulations is that of making the chamber volume *increase* initially, before a rapid compression just prior to discharge. This allows the compressor to use the leakage in its favour by sucking more gas in from the atmosphere.

- Is it physically possible to design a scroll for which the chamber volume varies nonmonotonically with time?

Many potentially important physical effects have been neglected in this study, and might be considered in the future to refine the model further.

- What would be the effects on our model of including (a) thermal losses through the compressor walls; (b) dependence of viscosity upon temperature; (c) the "squeeze film" and "Couette" effects in the lubrication analysis (i.e. the terms multiplied by Ω in (1.25, 1.26))?

Perhaps most importantly, we have restricted our analysis to a two-dimensional configuration.

- How could one quantify leakage through the *sides* of the chambers (i.e. in the plane shown in figure 1.2)?

Finally let us return to the three original questions posed in the introduction and examine the extent to which they have been answered. Although the first specific question posed has not been addressed in this case study, some progress was made at the Study Group, and in [2] a simple geometric method is given to parametrise a large class of viable compressor spirals. Our modelling provides a framework within which the second specific question from the introduction can be answered, since different proposed designs can be simulated and their performance and efficiency compared. Moreover, we have proposed some general design strategies which might enhance performance. Finally, in answer to the third question from the introduction, we have identified an easily evaluated functional which measures leakage and could form part of a cost function.

Acknowledgements

I am very grateful to Stig Helmer Jørgensen (DANFOSS) for stimulating my interest in the problem and to the other participants in the Study Group, in particular Jens Gravesen and Christian Henriksen, for useful discussions. I also received some helpful advice from Hilary and John Ockendon.

References

[1]   Creux, L. (1905) *Rotary Engine*. US Patent 801182.
[2]   Gravesen, J., Henriksen, C. & Howell, P. D. (1998) DANFOSS: Scroll optimization. In *Proc. 32nd European Study Group with Industry* (eds. Gravesen, J. & Hjorth, P. G.), Technical University of Denmark, 3–35.
[3]   McCullough, J. E. & Hirschfeld, F. (1979) The scroll machine — an old principle with a new twist. Mech. Eng. **101**, 46–51.

[4] Ockendon, H. & Ockendon, J. R. (1995) *Viscous Flow*, Cambridge University Press.

[5] Van Wylen, G. J., Sonntag, R. E. & Borgnakke, C. (1994) *Fundamentals of Classical Thermodynamics*, Wiley, New York.

Peter D. Howell

School of Mathematical Sciences, University of Nottingham,
University Park, Nottingham NG7 2RD, UK
Peter.Howell@nottingham.ac.uk

2

Determining the Viscosity of a Carbon Paste Used in Smelting

Preface

In the following case study, the slow viscous flow of blocks of "carbon paste" is analysed. The paste blocks are essential components of an electric smelting process by which a variety of ferro-alloys and other substances are produced. The problem is first proposed in its most general form. A nondimensionalisation using typical parameter values of the process then shows that a much simpler set of equations may be used to analyse the flow. After examining the qualitative details of the fluid motion in various key regions of the flow, an asymptotic analysis of a long thin block of paste allows us to develop a good general understanding of the main principles of slow viscous flow in paste blocks. The theory highlights the key differences between various tests that are used to determine the viscosity of carbon paste. Finally, some fairly straightforward numerical analysis is used to show that the general problem is amenable to simple methods; the numerical results also show that in many cases the "long thin" analysis used earlier can produce remarkably accurate results.

The work outlined below is part of research that was produced during and after the 1988 European Study Group with Industry, which was held at the University of Heriot-Watt, Scotland. Some extensions to the work presented here are suggested as projects for the interested student in section 2.6. Close links have continued to be maintained between industrial mathematicians and the ELKEM ASA, the Norwegian company that originally proposed the problem. A number of other problems (see [1], [4], and [3]) concerning various aspects of the electric smelting industry have also been considered in detail and have led to some interesting mathematical problems.

2.1 Continuous Electrode Smelting

Electric smelting is a popular process for producing a variety of materials such as ferro-alloys, silicon, and calcium carbide. Heat is provided for the

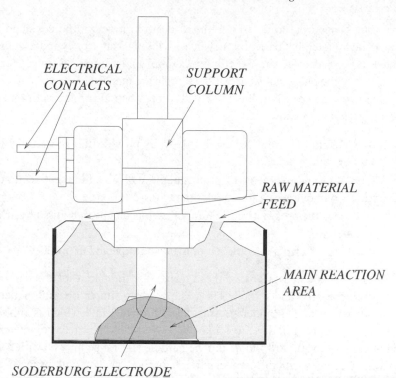

ELECTRICAL CONTACTS

SUPPORT COLUMN

RAW MATERIAL FEED

MAIN REACTION AREA

SODERBURG ELECTRODE

Fig. 2.1. Schematic diagram of carbon electrode furnace

smelting furnace via substantial (up to 150,000 Amps) electrical currents that are passed to the heart of the furnace through a carbon electrode. Figure 2.1 gives the schematic details of the physical arrangement.

As the smelting takes place, the electrode is consumed (depletion rates of 1m per day are typical) and must be replenished. A practical way of achieving this is to produce a continuous carbon electrode immediately above the furnace. This is normally done by feeding cylinders of carbon "paste" into the centre of a cylindrical steel casing (diameter 1–2 m) which is gradually heated. As the paste warms, it flows to fill the cylinder and is eventually "baked" solid at around 500°C in the region where the current enters the electrode. Baking improves both the strength and electrical conductivity of the electrode, and is essential for satisfactory current transfer. Electrodes of this type have come to be called Søderburg electrodes after the Norwegian who pioneered their development in the early part of this century.

To design an efficient Søderburg electrode, the viscosity of the carbon "paste" must be known, and this will be the main question that concerns us here. Traditionally at the smelting factory a number of simple experiments have been carried out to determine the viscosity. These will be considered more in detail later, but basically they involve taking a cube or cylinder of the heated carbon paste and simply letting it "slump" under its own weight. Measurements of the bulge at the base of the sample are made, and the viscosity is inferred from the growth of the bulge with time. The industrial practitioners specifically want to know:

- Which sort of viscosity test is likely to provide the most reliable answers?
- What size and geometry of paste sample will give the most accurate results without taking too much time to test?
- How does the prospective duration of each test depend on the physical parameters?
- What sort of errors may the experiments be expected to produce?

Before we can answer any of these questions, the nature of the carbon paste itself must be understood. The paste is not a simple one-component fluid, but consists of a binder (normally pitch or tar) into which is mixed particles of calcined anthracite (coke). The particles may have diameters ranging from 125 μm to 15 mm, and the particle size range may be varied to produce electrodes with different electrical, thermal, and strength properties. Experience has shown, however, that to all intents and purposes the paste may be taken to behave as a single viscous fluid. It is worth mentioning that there *are* circumstances for which this is not true, as the paste may "segregate" in the electrode. During segregation the particles clump together and large regions of the paste consist only of binder. Though this may cause severe problems during the smelting process, it is not our concern here; for more details concerning this interesting subproblem, the reader is referred to [1]. Regarding the thermal properties of the paste (which is solid at room temperature), experiments have shown that, at 50–80°C softening takes place and flow occurs. When the viscosity is being measured experimentally, tests are carried out at a constant temperature. This is convenient for our purposes, since it means that the thermal properties of the paste need enter into the problem only as parameters.

As far as the viscosity experiments are concerned, there are three recognised ways of causing a heated sample to "bulge" to allow a viscosity measurement to be made. These are:

(1)   The "plasticity" test: a sample of paste is heated to around 300°C, placed on a rigid impermeable surface, and allowed to slump under its own weight.

(2) The "velocity test": a metal plate is applied to the top (planar) surface of the sample and forced to move downwards with a (normally constant) prescribed velocity.

(3) The "viscometer": identical to the velocity test except that a constant force is used.

The terminology is used in deference to that standard in the industry. In fact all three tests are really viscometers, and no plastic flow in the normal solid-mechanics sense of the word occurs in the plasticity test. It is also worth noting that the velocity test and the viscometer test generally take much less time to complete than the plasticity test.

2.2 Problem Formulation

We fix our ideas by considering a block of paste occupying a region D which is initially rectangular with height h and semiwidth L. As shown in figure 2.2, the top surface of the sample is denoted by S_T, the surface in contact with the rigid plane at $y = 0$ by S_B and the (initially vertical) sides of the sample by S_- and S_+.

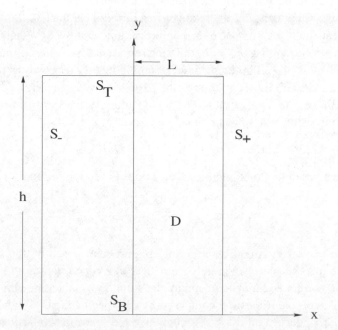

Fig. 2.2. Geometry and nomenclature of paste block

The first priority in the analysis of the problem is to determine which parameters are important and which effects dominate, and we tackle this by nondimensionalising. We carry this out by scaling lengths with h (thus tacitly assuming that the paste block has an aspect ratio of order 1), the fluid velocity q with U_∞ where U_∞ is a representative speed (for example the average speed of the centre of the top of the sample), the time t with a representative time τ, and the pressure p and the stress tensor T with $\mu U_\infty / h$, where μ is the dynamic viscosity (to be determined). The Navier–Stokes equations (see, for example [7]) with a gravity body force then become, in nondimensional variables,

$$\text{Re}[\text{St}^{-1} q_t + (q \cdot \nabla)q] = -\nabla p + \nabla^2 q - (\text{Re}/\text{Fr}) j,$$

$$\nabla \cdot q = 0,$$

where j denotes the unit vector in the y direction. Here the subscripts denote differentiation, and as usual the Reynolds, Froude, and Strouhal numbers are defined by

$$\text{Re} = \frac{hU_\infty \rho}{\mu}, \qquad \text{Fr} = \frac{U_\infty^2}{gh}, \qquad \text{St} = \frac{\tau U_\infty}{h},$$

where ρ denotes the fluid density (assumed known).

We must now consider the relative sizes of these parameters. Since none of the tests involves impulsive loading, we assume that the Strouhal number is of order one, so that $U_\infty \sim h/\tau$. Using representative values of $h \sim 1$ m, $U_\infty \sim 1$ m/h, $\rho \sim 3000$ kg/m^3, and assuming that the viscosity is of order of magnitude 10^8 Pa sec (although the paste viscosity may vary by orders of magnitude, the representatives from the factory assured us that this was a typical viscosity value), we find that

$$\text{Re} = O(10^{-8}), \qquad \text{Fr} = O(10^{-8}).$$

To lowest order, the nondimensional equations of motion become

$$\nabla p = \nabla^2 q - \alpha j \qquad (2.1)$$

$$\nabla \cdot q = 0, \qquad (2.2)$$

where $\alpha = \text{Re}/\text{Fr}$. Already, the formidable problem posed by the Navier–Stokes equations has been greatly simplified; to a good order of approximation, the correct equations are simply the slow flow equations with a body force. The problem must be completed by the specification of suitable initial and boundary conditions. Assuming that the shape of the paste block is specified at $t = 0$, we have the standard no-slip viscous boundary condition on $y = 0$, the sides S_+ and S_- are stress-free, and the boundary conditions that must be applied on the top surface S_T depend on which test we are using.

The equations of motion (2.1) and (2.2) hold for arbitrary paste blocks, but may be further simplified if the pressure is eliminated and we work only in *two dimensions*. Taking the curl of (2.1) and defining a stream function $\psi(x, y)$ so that $u = \psi_y$ and $v = -\psi_x$, the full problem becomes

$$\nabla^4 \psi = 0 \quad ((x, y) \in D),$$

$$\psi = \psi_y = 0 \quad ((x, y) \in S_B),$$

$$T \cdot n = 0 \quad ((x, y) \in S_+ \cup S_-),$$

$$(D/Dt)[x - \xi(y, t)] = 0 \quad ((x, y) \in S_+ \cup S_-), \tag{2.3}$$

$$\left.\begin{array}{lll} T \cdot n = 0 & & \text{(plasticity test)} \\ \psi_y = 0 & \psi_x = \dot{s}(t) & \text{(velocity test)} \\ \psi_y = 0 & -p - \psi_{xy} = \gamma(t) & \text{(viscometer)} \end{array}\right\} \quad ((x, y) \in S_T).$$

Here $D/Dt = \partial/\partial t + (q \cdot \nabla)$ denotes the standard convective derivative, n is the unit normal to the boundary, and $\dot{s}(t)$ and $\gamma(t)$ denote a nondimensional speed and normal stress respectively. The boundary of the sample is denoted by $x = \xi(y, t)$. As usual in slow flow problems, time enters into the problem only as a parameter and via the boundary conditions; the body force enters only via the pressure p in the boundary conditions.

2.3 Simplified Analysis

The full problem in the previous section is too complicated to solve in closed form, and a numerical solution is required. For the reasons explained above, however, we first search for some simplified versions of the problem where theoretical progress may be made.

2.3.1 Corner Solutions

First, we consider the general form of solutions near to a corner. Although such local analysis is unlikely to answer any of the industrialist's questions directly, at this stage any information on the nature of the flow is helpful. Each test may be examined, but, confining ourselves to the plasticity test and leaving the other cases as exercises for the interested student, it is easy to show that the local behaviour of solutions may be determined by simple separation of variables. Using x and y to denote local coordinates, we find that:

(i) near a top corner (where two stress-free surfaces meet),

$$u \sim -\frac{1}{4}\alpha xy, \quad v \sim \frac{1}{8}\alpha(x^2 + y^2), \quad p \sim -\frac{1}{2}\alpha y,$$

(ii) near the line of symmetry at the top,

$$u \sim \alpha\beta xy, \quad v \sim -\frac{1}{2}\alpha\beta(x^2 + y^2), \quad p \sim \alpha(-1 - 2\beta)y,$$

where β is an arbitrary constant that would have to be determined by a matching procedure (see, for example [6]).

(iii) near a bottom corner (where a stress free surface meets a fixed surface),

$$u \sim -\frac{1}{3}\alpha xy, \quad v \sim \frac{1}{6}\alpha y^2, \quad p \sim -\frac{2}{3}\alpha y,$$

(iv) near the bottom line of symmetry,

$$u \sim \alpha\delta xy, \quad v \sim -\frac{1}{2}\alpha\delta y^2, \quad p \sim \alpha(-1 - \delta)y,$$

where δ is arbitrary. Although these expressions for u, v, and p may easily be verified to be solutions to the local problem, evidently they all have the property that $u = 0$ at $x = 0$, and thus predict no bulging of the sample. To produce the required outward movement of the walls of the sample, it is necessary to recognise that the flow is dominated by an eigensolution. The eigenproblem is simple to pose: near the bottom left-hand corner, for example, employing local polar coordinates, the equation of motion is

$$\nabla^4 \psi = 0 \quad (r \geq 0, \ 0 \leq \theta \leq \pi/2),$$

while the boundary conditions are the no-slip conditions

$$\psi_r = \psi_\theta = 0 \quad (\theta = 0)$$

and the zero-stress conditions on the free boundary $\theta = \pi/2$. To express these in a convenient form, we note that the stress tensor is given (in dimensional variables) by

$$T = \begin{pmatrix} -p + 2\mu u_x & \mu(u_y + v_x) \\ \mu(u_y + v_x) & -p + 2\mu v_y \end{pmatrix},$$

and since the outward normal to the free boundary is given by $n = (-1, 0)$, the stress-free condition amounts to (returning to nondimensional variables)

$$-p + 2u_x = u_y + v_x = 0 \quad (\theta = \pi/2).$$

The second condition may easily be transformed to polar coordinates to give

$$r^2 \psi_{rr} - r\psi_r - \psi_{\theta\theta} = 0,$$

while the first may be dealt with by differentiating with respect to y and using the y-momentum equation to eliminate the term p_y. This finally leads to

$$\psi_{\theta\theta\theta} - 3r\psi_{r\theta} + 4\psi_\theta + 3r^2\psi_{rr\theta} = 0 \quad (\theta = \pi/2),$$

which is the final boundary condition for the homogeneous problem.

Having posed the problem for the bottom corner, the top corner may be examined in a similar way, giving the problem

$$\nabla^4 \psi = 0 \quad (r \geq 0, \ -\pi/2 \leq \theta \leq 0),$$

$$\left.\begin{array}{l} r^2\psi_{rr} - r\psi_r - \psi_{\theta\theta} = 0 \\ \psi_{\theta\theta\theta} - 3r\psi_{r\theta} + 4\psi_\theta + 3r^2\psi_{rr\theta} = 0 \end{array}\right\} \quad (\theta = -\pi/2 \ \text{and} \ \theta = 0),$$

Both problems may be solved by standard separation of variables, or more simply by seeking a solution of the form

$$\psi = r^\lambda [A\cos\lambda\theta + B\sin\lambda\theta + C\cos(\lambda-2)\theta + D\sin(\lambda-2)\theta],$$

where λ, A, B, C, and D are constants. Imposing the boundary conditions, we find that for the bottom corner either $\lambda = 1$ (giving again that u is zero along $\theta = \pi/2$) or λ must satisfy the equation

$$\tan^2(\pi\lambda/2) = (\lambda-1)^2/\lambda(2-\lambda). \tag{2.4}$$

It may easily be shown that (2.4) has precisely two real solutions $\lambda \sim 0.405$ (which must be rejected since it leads to singular behaviour for u at $r = 0$) and $\lambda \sim 1.595$. Thus an eigensolution with a velocity

$$\psi_r \sim r^{0.6}$$

along $\theta = \pi/2$ is possible. This suggests that the bulging that takes place in the plasticity test is associated with a free boundary that grows like $tr^{0.6}$. Near to the top corner, λ must satisfy

$$\tan^2(\pi\lambda/2) = \lambda(\lambda-2)/(1-\lambda)^2, \tag{2.5}$$

an equation whose only real zeros are 0, 1, and 2. All three of these solutions give $u = 0$ on $\theta = -\pi/2$ and therefore do not allow bulging. (It is worth pointing out that (2.4) and (2.5) possess infinitely many complex solutions as well as the real ones discussed above. These have not been investigated, but almost certainly correspond to the classical "Moffatt vortex" solutions (see for example [5]) that are typical in slow flow problems).

The conclusions of our local analysis may be summarised as follows.

(i) The behaviour at the corners is dominated by eigensolutions.

(ii) With or without eigensolutions, the bottom corner point cannot move. This is essentially a consequence of the no-slip condition and is in direct contrast to the inviscid case. (For analysis of the inviscid case see [8].)

(iii) When eigensolutions are included, bulging can take place at the bottom of the sample but not at the top.

2.4 Special Geometries

Even the simple corner solutions examined in the previous section have allowed us to begin to piece together what happens to a sample of paste during testing. The top surface initially remains much as it began, and most of the movement occurs at the bottom of the sample, since bulging occurs at the corners. To further analyse the process, we consider some special geometries. Though it is tempting to examine the case of a "short fat lump of paste", it may easily be shown that little of value can be wrung from this case. Though the geometry leads to some simplifications, it is clear from the outset that there will be boundary layers at the edges of the sample, and these are precisely the regions that interest us most. Analysis of the boundary layers involves consideration of the original full problem, and thus little progress may be made.

The prognosis is altogether more encouraging if we consider the other obvious limit of a tall thin sample of paste. Assuming that $L/h = \epsilon \ll 1$, we consider the region $x > 0$ for simplicity and impose symmetry conditions on $x = 0$ (which will subsequently be denoted by S_A). Making the obvious scalings $x = \epsilon X$, $u = \epsilon U$, and $\xi = \epsilon \eta$, the scaled nondimensional problem becomes

$$p_X = U_{XX} + \epsilon^2 U_{yy}, \tag{2.6}$$

$$\epsilon^2 p_y = v_{XX} + \epsilon^2 u_{yy} - \epsilon^2 \alpha, \tag{2.7}$$

$$U_X + v_y = 0. \tag{2.8}$$

Assuming that $U = U_0 + \epsilon U_1 + \epsilon^2 U_2 + \cdots$ and using similar expansions for v, p, and η, a regular perturbation solution of (2.6)–(2.8) that additionally satisfies the symmetry conditions $U = v_X = 0$ on $X = 0$ may easily be determined. Correct to order ϵ^2, we find that

$$U = -Xv_{0y} - \epsilon Xv_{1y} + \epsilon^2 \left[-\frac{1}{6}X^3(-2v_{0yyy} + g_{0yy}) - Xq_y \right],$$

$$v = v_0(y, t) - \epsilon v_1(y, t) + \epsilon^2 \left[\frac{1}{2}X^2(g_{0y} + \alpha - 2v_{0yy}) + q(y, t) \right],$$

$$p = -(v_{0y} + g_0) + \epsilon(-v_{1y} + g_1) + \epsilon^2 \left[\frac{1}{2}X^2(v_{0yyy} - g_{0yy}) + g_2(y, t) \right],$$

where the functions q, g_1, g_2, and g_3 remain to be determined. Imposing the free surface kinematic boundary condition gives

$$(\eta_0)_t + (\eta_0 v_0)_y = 0 \tag{2.9}$$

to leading order, and another equation relating v_0 and η_0 may be determined by imposing the stress-free conditions on $X = \eta(y, t)$. There are two components of $T \cdot n$ that must be zero; the x component gives

$$-g_0 + v_{0y} = -g_1 - v_{1y} = 0$$

correct to $O(\epsilon)$, while the $O(\epsilon^2)$ contribution relates the functions q, g_2, and g_3. The y component yields nothing until we reach the $O(\epsilon^2)$ terms, whereupon we find that

$$4(\eta_0 v_{0y})_y = \alpha \eta_0. \tag{2.10}$$

The leading-order solution thus becomes

$$U = -Xv_{0y} - \epsilon Xv_{1y} + \epsilon^2 \left[\frac{1}{2}X^3(v_{0yyy} - Xq_y \right],$$

$$v = v_0 - \epsilon v_1 + \epsilon^2 \left[\frac{1}{2}X^2(\alpha - 3v_{0yy}) + q \right],$$

$$p = -2v_{0y} - 2\epsilon v_{1y} + \epsilon^2 \left[X^2 v_{0yyy} + g_2 \right],$$

where η_0 and v_0 are determined by (2.9) and (2.10). As far as boundary and initial conditions for the equations for η_0 and v_0 are concerned, clearly it makes sense to specify an initial condition for η_0. Also, since (2.10) is of second order in y, we expect to be able to satisfy two boundary conditions. The full problem requires that two stress conditions are satisfied on the top surface of the sample, while two no-slip conditions apply to the bottom of the sample. Therefore two conditions must be dropped, and we may expect there to be boundary layers at the top and bottom of the sample. Since we want results that reflect properties of both the top and the bottom of the sample, we impose $v_0 = 0$ on $y = 0$ and one stress condition on the sample top.

2.4.1 Further Analysis of the Velocity Test

Now that the framework for analysing the long thin paste block has been established, we derive some results. For the velocity test, where v is specified on the top of the sample, some progress may be made with the governing equations for the period soon after the experiment begins. Expanding v_0 and η_0 in powers of t, assuming that $v = -V_T$ on $y = 1$ and $\eta_0(y, t) = 0$ (an initially straight sided block) the solution satisfying $v_0 = 0$ on $y = 0$ is given by

$$\eta_0 = 1 + \frac{t}{8}(8V_T - 2\alpha y + \alpha) + O(t^2),$$

$$v_0 = \frac{1}{8}\alpha y^2 - y\left(V_T + \frac{\alpha}{8}\right)$$
$$+ \frac{t}{192}\left[\alpha y(2\alpha y^2 - 3\alpha y + \alpha - 24V_T y + 24V_T)\right] + O(t^2).$$

Three separate cases are identified (as shown in figure 2.3):

(i) For $V_T > \alpha/8$, both the top and the bottom of the sample will bulge as the top plate is pushed down. The free boundary is linear in t, and the horizontal velocity on the top and the bottom of the sample is proportional to x.

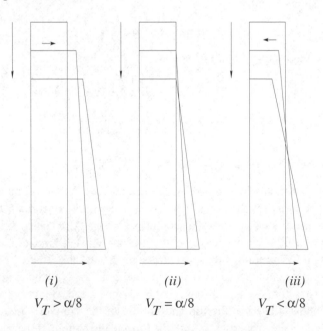

(i)   (ii)   (iii)

$V_T > \alpha/8$ $V_T = \alpha/8$ $V_T < \alpha/8$

Fig. 2.3. Schematic paste block shapes for different V_T

(ii) When $V_T = \alpha/8$, the imposed velocity is just large enough to ensure that $\eta_0 = 1$ at $y = 1$, so that the horizontal velocity on the top surface is zero.

(iii) When $V_T < \alpha/8$, the sample "necks" as the top surface moves towards the y axis, while the bottom surface moves away from it.

Near $y = 0$ and $y = s(t)$ the analysis is not valid, since boundary layers are present; nevertheless the outer solution may still be used to predict the viscosity from the bulging. Evaluating η_0 on $y = 0$ and restoring the dimensional variables reveals that, if the top plate moves with speed U_∞ and the maximum semiwidth of the sample at any time is BL, say, then the dynamic viscosity is given by

$$\mu = \frac{h^2 \rho g t}{8(h(B-1) - U_\infty t)}. \tag{2.11}$$

2.4.2 Analysis of the Viscometer Test

The only change required to examine this case for small t is to impose the boundary condition $p_0 + 2v_{0y} = \gamma(t)$ as well as $v_0 = 0$ on $y = 0$. For a constant load $\gamma(t) = -\gamma$, say, we then easily find that

$$\eta_0 = 1 + \frac{t}{4}(\gamma - \alpha y + \alpha) + O(t^2),$$

$$v_0 = \frac{1}{8}\alpha y^2 - \frac{y}{4}(\gamma + \alpha)$$
$$+ \frac{t}{96}[\alpha y(\alpha y^2 - 3\alpha y + 3\alpha - 3\gamma y + 6\gamma)] + O(t^2).$$

The predicted bulge is therefore once again linear in y, though no necking of the sample can occur unless $\gamma < 0$. (Although such an experiment would be rather hard to set up, there is no reason why the viscosity could not be measured by forcing the top of the sample to rise and measuring the necking.) For a maximum sample semiwidth BL and a (dimensional) load Γ, redimensionalisation gives the result corresponding to (2.11) as

$$\mu = \frac{t(\Gamma + h\rho g)}{4(B-1)}. \tag{2.12}$$

Actually, some extra information is available here, since in this case the speed of the top of the sample is also unknown. The possibility thus arises of estimating the viscosity from the speed of the plate. Assuming that $\eta_0 = 1 + O(t)$ for $t \ll 1$ and solving (2.9) and (2.10) for small time, we find, on imposing the conditions that $v_0 = 0$ on $y = 0$ and $-p_0 + 2v_{0y} = -\gamma$ on $y = s(t)$, that

$$v_0 = \frac{1}{8}\alpha y^2 + \frac{y}{4}[-\gamma - \alpha s(t)].$$

Insisting that $v_0 = \dot{s}$ on $y = s(t)$ thus gives an ordinary differential equation for $s(t)$, which has the solution

$$s(t) = \frac{2\gamma e^{-\gamma t/4}}{\alpha + 2\gamma - \alpha e^{-\gamma t/4}}.$$

A redimensionalisation now shows that the effective viscosity is given in terms of s by

$$\mu = \frac{\Gamma t}{4\log[(2h\Gamma + h\rho gs)/s(h\rho g + 2\Gamma)]} \sim \frac{ht(\Gamma + h\rho g/2)}{4(h - s)}.$$

2.4.3 Analysis of the Plasticity Test

The plasticity test involves a further complication: there is no reason why the top surface of the sample should remain flat. However, it is fairly easy to show that, to leading order, the top surface *does* remain parallel to the x axis. Small-time expansions then show that

$$\eta_0 = 1 + \frac{t\alpha}{4}(1 - y) + \frac{\alpha t^2}{32}(-2\alpha y + \alpha y^2 - 4D) + O(t^3),$$

$$v_0 = \frac{1}{8}\alpha y^2 - \frac{\alpha y}{4} + \frac{ty}{96}[\alpha^2 - 3\alpha^2 y + \alpha^2 y^2 + 24\alpha D)] + O(t^2),$$

where D is a constant that could be obtained by going to higher order. For, therefore, the viscosity may be estimated using

$$\mu = \frac{h\rho g t}{4(B - 1)}. \tag{2.13}$$

2.4.4 The Boundary Layer at the Base of the Sample

The results (2.11), (2.12), and (2.13) give us our first concrete formulae, and, as we shall see later, are capable of giving accurate predictions. Before proceeding, however, some account must be taken of the boundary layers at the top and bottom of the sample. Although for practical purposes it is probably not necessary to give all the details of the boundary layer problem, good mathematical practice demands that we should at least satisfy ourselves that the problem is well-posed. It is evident that in both boundary layers the full problem applies, and thus a numerical solution is required in general. Some progress may be made, however, with the small-time problem. Considering by way of an example the boundary layer at the base of the sample in the velocity test, we exploit the facts that the boundary S_+ remains vertical for $t \ll 1$ and the outer velocity is known to be

$$v_0 = -X \left[\frac{1}{8} \alpha y^2 + K y \right] + O(t) \quad (K = -V_T - \alpha/8).$$

Introducing a stream function Ψ that has been scaled (in addition to y) with ϵ and further setting $\Psi = -KYX + \Phi$, the matching problem becomes, to leading order,

$$\nabla^4 \Phi = 0 \quad (0 \le X \le 1, \ 0 \le Y < \infty)$$

with boundary conditions

$$\Phi = \Phi_{XX} \quad (X = 0),$$

$$\Phi_{YY} - \Phi_{XX} = 0, \quad \Phi_{XXX} + 3\Phi_{XYY} = 0 \quad (X = 1),$$

$$\Phi = 0, \quad \Phi_Y = KX \quad (Y = 0)$$

and matching condition $\Phi \to 0$ as $Y \to \infty$. A symmetrical separation of variables solution that is suitably behaved as $Y \to \infty$ is given by

$$\Phi = \exp(-\lambda Y)[BX \cos \lambda X + C \sin \lambda X],$$

while the two conditions on $X = 1$ mean that there can only be non-trivial solutions for the constants B and C if

$$\lambda + \sin \lambda \cos \lambda = 0. \tag{2.14}$$

The expression (2.14) is familiar in both elasticity and slow viscous flow; it is the Papkovitch–Fadle equation, and many of its properties are known. In particular it possesses an infinite number of complex solutions which are given, for large k, by

$$\lambda_k = \left(k - \frac{1}{2} \right) \pi + \frac{1}{2} i \log(4k\pi) + O(\log k/k).$$

Although it is possible to completely determine the coefficients of the eigenfunction expansion so that the conditions at $Y = 0$ are satisfied, it is not a trivial matter, since the problem is a so-called "non-canonical" one. Biorthogonal functions and collocation must be used, the net result being an infinite system of linear equations for the coefficients, which may be made diagonally dominant provided sufficient care is taken. Of most importance is the fact that, though the minutiae are complicated (see, for example [9]), the details may all be taken care of.

2.5 Numerical Analysis and Results

The analysis that was carried out on the Navier–Stokes equations allows the numerical problem for general blocks of paste to be considerably simplified. Instead of solving a fully time-dependent nonlinear problem, we now have to solve what amounts to a fourth-order linear time-independent equation of motion. The position of the free surface may then be advanced using the kinematic condition. Space does not permit any but the barest numerical details (a full discussion may be found in [2]), but essentially the most convenient way to solve the problem is to use the finite-element method to solve the partial differential equation, followed by a Lagrangian method to advance the boundary.

2.5.1 Finite-Element Method

To apply the finite-element method, the equations are most conveniently written as

$$\sigma_{11x} + \sigma_{12y} = 0, \qquad \sigma_{21x} + \sigma_{22y} = 0, \qquad u_x + v_y = 0, \qquad (2.15)$$

where

$$\sigma_{11} = -p + 2u_x, \qquad \sigma_{12} = \sigma_{21} = u_y + v_x, \qquad \sigma_{22} = -p + 2v_y.$$

The boundary conditions are

$$u = v = 0 \ ((x, y) \in S_B), \quad T_x = T_y = 0 \ ((x, y) \in S_+),$$

$$u = T_y = 0 \ ((x, y) \in S_A),$$

and, for $(x, y) \in S_T$,

$$T_x = T_y = 0 \qquad \text{plasticity test}$$

$$u = 0, \ v = \dot{s}(t) \qquad \text{velocity test}$$

$$u = 0, \ T_y = \gamma(t) \qquad \text{viscometer test},$$

where the surface tractions T_1 and T_2 are given by

$$T_1 = \sigma_{11}\frac{\partial x}{\partial n} + \sigma_{12}\frac{\partial y}{\partial n}, \qquad T_2 = \sigma_{12}\frac{\partial x}{\partial n} + \sigma_{22}\frac{\partial y}{\partial n},$$

n being in a direction normal to the boundary. Dividing the flow region D into triangular elements with corner and midside nodes, the velocity components are approximated by polynomials one degree higher than those used for the pressure. Using quadratic and linear polynomials respectively, we choose the coefficients of the polynomials to satisfy the weak (integrated) forms of (2.15). (Details of the numerical procedures may be found in standard text books on finite and boundary elements.) In this way the whole problem may be reduced to a system of linear equations which may then be solved by standard methods. Once the solution has been computed at a particular time, the Lagrangian equations

$$\frac{dx}{dt} = u, \quad \frac{dy}{dt} = v$$

are integrated in an explicitly discretised form to advance the boundary, whereupon the whole process begins again. The scheme constitutes a quick and accurate method of solving the full problem. Although for simplicity only two-dimensional cases are considered here, the method could also be extended to three dimensions with little extra effort.

2.5.2 Results

Numerical results are shown for a typical paste block of unit aspect ratio in figure 2.4. The calculations were performed using 100 elements (giving 231 nodes and 528 variables), and numerical tests suggested that this gave results with a relative error of about 10^{-4}. The results display qualitative features that are familiar to Elkem from their experiments: in the plasticity test there is a pronounced bulge near to the bottom of the sample, whilst the bulging is more evenly distributed along the lateral boundary of the sample in the viscometer. In both the plasticity test and the viscometer, the top edge of the boundary nearest to the free surface tends to become slightly elevated. This cannot of course happen in the velocity test, where once again the bulging seems to take place over most of the free boundary rather than in a local area. It is of interest to compare the numerically calculated results to the predictions of (2.11), (2.12), and (2.13). In all of the results given below, the paste sample had a viscosity of $\mu_p = 10^8$ Pa sec and a density of 3000 kg/m^3. The gravitational constant g was taken to be 9.8 m/s^2, the paste block was assumed to be 1 m high, and the velocity scale was taken to be 1 m/hr. For the velocity and viscometer tests, V_1 and γ were both taken to be 1. Table 2.1 shows results for the paste block shown in figure 2.4. The time increment is denoted by n, which varied for different tests; this reflects the fact that each method of viscosity measurement has its own characteristic experimental time. For the results discussed below, it was found appropriate to use time increments

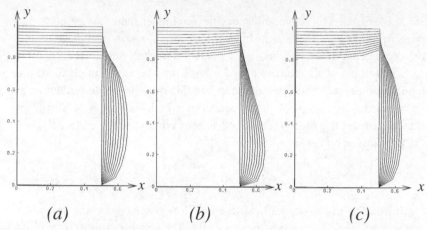

(a)   *(b)*   *(c)*

Fig. 2.4. Numerical results for paste block of unit aspect ratio (100 elements) (a) viscosity test, (b) plasticity test, (c) viscometer

Table 2.1. *Viscosity Ratios* μ/μ_p *Predicted by Asymptotic Model in the Case* $\epsilon = 1$

n	Velocity Test (Bl)	Plasticity Test (Bl)	Plasticity Test (Ht)	Viscometer (Bl)	Viscometer (Ht)
1	0.436	1.657	1.170	1.299	1.199
2	0.402	1.665	1.170	1.301	1.202
3	0.373	1.674	1.171	1.302	1.206
4	0.347	1.683	1.171	1.304	1.210
5	0.325	1.693	1.171	1.307	1.213
6	0.305	1.703	1.172	1.310	1.217
7	0.288	1.714	1.172	1.313	1.221
8	0.273	1.725	1.173	1.317	1.225
9	0.259	1.738	1.174	1.322	1.230
10	0.247	1.751	1.175	1.327	1.235

of 0.02 h, 0.2 h, and 0.07 h respectively for the velocity, plasticity, and viscometer tests. The table gives values of μ/μ_p (which should take the value 1 if the theory is perfect) for each experiment, μ having been calculated via the formula discussed previously. Results (labelled "Ht") are also given for cases where the height of the sample is measured instead of the bulge.

Table 2.1 compares exact results with a theory that is valid in the limit $\epsilon \to 0$ for the case $\epsilon = 1$, and so one would not expect there to be any agreement. Nevertheless, all of the results indicate that the predicted viscosities are at least of the correct order of magnitude. Although the velocity test predicts viscosities that are incorrect by factors of 2–4 (depending on when observations are made), the plasticity and viscometer tests give remarkably

Table 2.2. *Viscosity Ratios* μ/μ_p *Predicted by Asymptotic Model in the Case* $\epsilon = 1/10$

n	Velocity Test (Bl)	Plasticity Test (Bl)	Plasticity Test (Ht)	Viscometer (Bl)	Viscometer (Ht)
1	0.956	1.169	1.009	1.090	1.014
2	0.829	1.166	1.016	1.091	1.023
3	0.728	1.162	1.023	1.091	1.032
4	0.646	1.159	1.029	1.091	1.041
5	0.578	1.156	1.036	1.091	1.050
6	0.521	1.153	1.042	1.091	1.058
7	0.472	1.150	1.048	1.091	1.067
8	0.430	1.148	1.053	1.091	1.075
9	0.393	1.145	1.059	1.091	1.083
10	0.361	1.143	1.064	1.090	1.091

accurate results, especially when the height of the sample is measured rather than the bulge.

Table 2.2 gives results when the aspect ratio of the sample is 10, so that the parameter $\epsilon = 0.1$ is truly small. Here we may anticipate that the theory will be much more accurate, and this expectation is borne out. Once again, the velocity test seems to be the least accurate experiment, though for small times it gives acceptable results. Each of the other tests gives results that are accurate to a few percent, though once again it seems that it is slightly better to measure the height of the sample rather than the bulge.

2.6 Final Conclusions

What has our modelling achieved? We began this study with a number of clearly specified goals, so let us review these one by one.

First, we now have a much improved physical understanding of how the process works. The type of bulging that may be expected for each test, "necking" and other effects may now all be explained in terms of a clearly defined mathematical framework.

The key nondimensional parameters have been identified: the Reynolds number is so small that inertia cannot be of any importance, and the process is driven by the parameter

$$\frac{\text{Re}}{\text{Fr}} = \frac{gh^2\rho}{\mu U_\infty}.$$

By investigating obvious alternative nondimensionalisations using the velocity and loading of the top plate, timescales could be determined for each test (details left to the reader). In particular, the velocities and loadings that equalise the time taken for each experiment to be performed could be determined.

After examining the results for the tall thin paste block, the practitioner should have a good feel for how changes in the experiment are likely to be reflected in the results. Examining the general properties of the model in this simple case shows how the results are likely to change if given parameters are altered; the results also show that acceptable results may be obtained even when the theory is stretched beyond its reasonable limits.

Finally, the original numerical problem has been vastly simplified. The slow flow problem that must be solved is so much simpler than the full Navier–Stokes problem that three-dimensional time-dependent computations are a real possibility; in any case the theory generates a number of test cases that may be compared with numerical codes that may be developed.

Further extensions to the work presented above are left to the reader. Future projects could include:

- It was assumed above that all of the tests are carried out at a constant temperature. But there is also interest in the "unsteady" temperature problem where a paste sample is put in a hot oven and flows as it heats up. How would this change the full problem, and what simple cases could still be solved in closed form to provide test cases for the numerics?
- What happens for a cylindrical block of paste? Does the change to such a geometry alter the qualitative flow of the paste? (This might be possible since the sample now only has sharp-angled sides at its top and bottom.)
- Suppose that an "instant" test was required where the sample was given a impulse (hit with a hammer, for example). How would this change the full problem? Could this experimental methodology reasonably be expected to give accurate results more quickly than the standard tests?
- As discussed above, it is known that under some circumstances the paste "segregates" and may no longer be regarded as a uniform mixture of pitch and fines. Given an initial segregation distribution, how would the modelling be affected?

Acknowledgement

Thanks are due to Dr. Svenn-Anton Halvorsen (Elkem a/s, Kristiansand, Norway) who first brought this problem to our attention at the 1988 Study Group with Industry, Heriot-Watt University.

References

[1] Bergstrom, T., Cowley, S., Fowler, A. C. & Seward, P. E. (1989) Segregation of carbon paste in a smelting electrode. IMA J. Appl. Math. **43**, 83–99.

[2] Fitt, A. D. & Aitchison, J. M. (1993) Determining the effective viscosity of a carbon paste used for continuous electrode smelting. Fluid Dynamics Res. **11**, 37–59.

[3] Fitt, A. D & Howell P. D. (1997) The manufacture of continuous smelting electrodes using carbon paste briquettes. J. Eng. Math. **33**, 353–376.

[4] Hinch, J. (ed.) (1995) Temperature variations and control of a calciner, *Proc. 28th Study Group for Mathematics in Industry*, University of Cambridge, 1995.

[5] Moffatt, H. K. (1964) Viscous and resistive eddies near a sharp corner. J. Fluid. Mech. **18**, 1–18.

[6] Nayfeh, A. H. (1973) *Perturbation Methods*, John Wiley & Sons.

[7] Ockendon, H. & Ockendon, J. R. (1995) *Viscous Flow*, Cambridge University Press.

[8] Penney, W. G. & Thornhill, C. K. (1952) The dispersion, under gravity, of a column of fluid supported on a rigid, horizontal plane. Phil. Trans. Roy. Soc. A **244**, 285–311.

[9] Spence, D. A. (1982) A note on the eigenfunction expansion for the elastic strip. SIAM J. Appl. Math. **42**, 155–173.

Alistair D. Fitt

Faculty of Mathematical Studies, University of Southampton, Southampton SO17 1BJ, UK

adf@maths.soton.ac.uk

3

The Vibrating Element Densitometer

Preface

The following case study concerns the resonant oscillations of an elastic plate in a fluid. At first sight one might expect that any analysis would get difficult quite quickly, forcing numerical solutions. However, simple beam theory suffices for the motion of the plate since it is thin, and potential theory for the motion of the fluid suffices since little viscous shear is generated. Moreover, amplitudes are very small, linear theory works, and a very simple solution generates the form of the answer that the client was seeking. The numerical results for this solution do not fit data at all well, but re-examination of the device indicates that a change in the boundary condition is required. This modification then provides results in remarkable agreement with data. The latter instructs on the importance of correctly modelling boundary conditions.

What follows is part of the analysis generated in projects completed by Claremont Mathematics Clinic teams for ITT-Barton, a company manufacturing the densitometer. The basic analysis was completed in 1982–3; various extensions were analyzed in 1983–4, see [8], [4]. A formal presentation was prepared, [2]. Only the basic analysis is presented here; extensions are suggested as projects for the interested student in section 3.7.

3.1 Introduction

As its name implies the densitometer is a device which measures density. (The reader is challenged to design a procedure which measures density to high accuracy – here the tolerance is 0.1% error – to operate safely in a hostile environment for long periods.) The device under consideration is used mainly in the oil industry, alongside volume flow rate devices, so that mass flow rates can be assessed in the transfer of energy products such as gasoline, kerosene, or heating oil. Energy content is proportional to mass, and so both density and volume flow rate are required for its correct quantification.

The device is illustrated in figure 3.1. The sensing head and probe are inserted into a hole of the fluid-carrying pipe, and the mounting flange is bolted to the pipe, with the electronic component housing exterior to the pipe. The sensing head is located at the center of the pipe's cross section. The sensing head is a cylinder open at both ends, its generators perpendicular to the flow direction. Fluid flows down the pipe around the sensing head cylinder, and flows through the cylinder only due to leakage and replacement, but this is considered sufficient to adequately measure any flow changes. A flat vane – a thin rectangular plate – is mounted across the central transverse plane of the mounting head cylinder. A cylinder connects the mounting flange to the sensing head. Inside this cylinder is a magnetostrictive driven core. An oscillating electric current fed to this core causes it to execute mechanical oscillations longitudinally, which, in turn, causes the vane to vibrate in a direction perpendicular to its plane. The vane is held along two opposing edges by the sensing head cylinder, but is thin and can flex in response to this external force. These vibrations are of extremely small amplitude; they are not apparent visually but can be felt at the vane. The sensing head is constructed from two cylinders, with the vane brazed into the inner cylinder, a piezoelectric crystal attached to the vane, and the outer cylinder enclosing this structure (shrunk-fit on). The piezoelectric crystal outputs an electrical current when subject to stress, and this current is fed to the outside of the pipe where it is recorded.

The densitometer device described above uses the phenomenon of resonance and exploits the fact that the resonant frequencies of a mechanical system will change if the system is immersed in fluids of different densities. If the plate is vibrated in a vacuum, it has one set of resonant frequencies, in air another set, in gasoline another set. In its vibration, the plate is pushing fluid away from itself above and below. This effect is similar to an added inertia of the plate proportional to some portion of local fluid and hence to fluid density. Analysis based on this concept – added mass – is presented in section 3.3.

In order to operate the device, it must be calibrated. The manufacturer was working with the relation

$$\rho_f = \frac{A}{\omega_n^2} - B \tag{3.1}$$

where ρ_f is the fluid density and ω_n is the natural frequency of the vane in that fluid. A and B are physical constants, which may vary from device to device. The particular values A and B have for a given device are found by operating the device in a number of fluids of known densities and recording

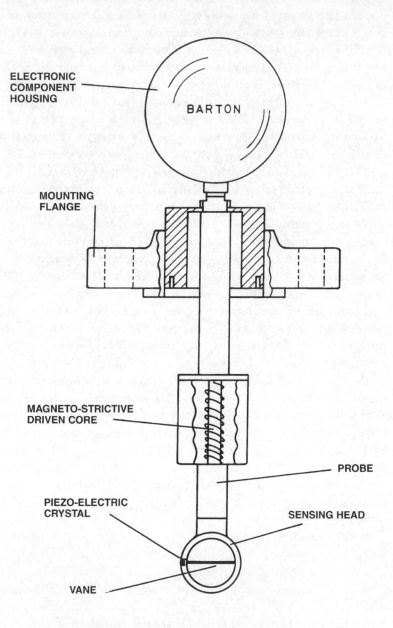

Fig. 3.1. Schematic diagram of densitometer device

the resonant frequencies. The optimum values of A and B which best fit these data are found by regression analysis (least squares error between the data and equation (3.1)). Equation (3.1) is programmed into the device electronics, which then outputs a density when a resonant frequency is identified.

The manufacturer came with a set of questions and goals, which included:

- What is the physical basis for equation (3.1)? This was not known to the current manufacturer, since the device had been passed on, company to company, in a series of acquisitions, and the original inventor is not now known.
- Is there a better formula than (3.1)?
- What are the physical constants involved in determining A and B from a realistic model?

With the information from the answers to the three questions posed above, the manufacturer might be able to improve quality control. A device is rejected for field work if, with the values of A and B found by regression, the data points lie outside a tolerance region around results from equation (3.1). A rejected device is disassembled, and its parts inspected and replaced if found defective. However, sometimes a device would not operate to the tolerance required even with no parts seen to be defective.

- What is the effect of temperature? The flange may be at an external temperature different from the fluid temperature.
- What other fluid properties affect the functioning of the device? In particular what are the roles of fluid viscosity and pressure?
- What is the role of fluid compressibility? The device specifications relate to liquids of specific gravity (ratio of liquid mass to mass of equal volume of water) range 0.3 to 1.2, but extrapolation of the range to gas densities is important.
- What is the effect of pre-stress? In the process by which the vane is attached to the inner cylinder, both are heated and joined by brazing. Since the cylinder and vane have different coefficients of thermal expansion the cooling may set up a residual pre-stress in the vane.

The above list does not exhaust the queries posed by the client, but it does give an indication of project goals. The main questions (the first three) were answered, and the modelling and analysis required for those results are reproduced here. Some of the remaining factors were explored successfully, some not successfully, and some were not attempted due to time constraints. The interested reader is encouraged to explore the modelling involved in extending the analysis to deal with the factors omitted from the basic model.

3.2 Resonance

This topic is quickly reviewed. Consider first a simple system which can undergo mechanical vibrations: a mass m is displaced $u(t)$ from its rest position and is acted on by an elastic restoring force ku, and a frictional (or damping) force opposing motion $r \frac{du}{dt}$. The governing equation of motion is

$$m \frac{d^2 u}{dt^2} + r \frac{du}{dt} + ku = A \cos \omega t \qquad (3.2)$$

where the right-hand side represents a periodic external force of frequency ω and amplitude A. In the absence of damping ($r = 0$) the particular solution of (3.2)

$$u = \begin{cases} \left[A / m (\omega_n^2 - \omega^2) \right] \cos \omega t , & \omega \neq \omega_n, \\ \left[A / 2 m \omega \right] t \sin \omega t , & \omega = \omega_n, \end{cases} \qquad (3.3)$$

where ω_n, the natural frequency of the system described by (3.2), is given by

$$\omega_n^2 = k / m.$$

The result (3.3) shows that the amplitude of the oscillation is constant for $\omega \neq \omega_n$, but becomes unbounded as $t \to \infty$ for $\omega = \omega_n$. This phenomenon is known as resonance, and is exploited in many mechanical (pushing a child on a swing) and electrical (tuning in a radio station) systems, and elsewhere. In the case of small damping, the amplitude is relatively large when ω is close to ω_n. (See [1].)

The above analysis applies to a "lumped parameter" system described by an ordinary differential equation. Such systems have a finite number of natural frequencies (a double pendulum, whose motion is described by two coupled second-order ODEs, has two natural frequencies). "Distributed parameter" systems, which are described by partial differential equations, as is the motion of the densitometer vane, typically have an infinite set of natural frequencies, corresponding to the infinite set of "modes" of their vibration. Consider plucking a guitar string of length L at locations $L/2, L/3, L/4 \cdots$ from one end.

We note that the set of natural frequencies is a parameter set identified with the system, independent of parameters defined by the input or forcing. The natural frequencies may be found analytically by the free oscillations of the system with no external forcing. Resonance occurs when there is an external agency providing an energy supply at a frequency coincident with one of the system's natural frequencies. So, in order to identify the natural frequencies of a system, one may apply a periodic external force and vary the input frequency until the output amplitude reaches a peak (or set of peaks). This is the technique used for the densitometer.

3.3 Added Mass Model

The simplest model for the motion of fluid is achieved by assuming the fluid to be incompressible, inviscid, and irrotational. The first assumption is highly reasonable for liquids, since large forces are required to change a liquid's density and here we are considering only extremely small vibrations. The second assumption is valid for fluid flow in which the effects of viscous shear are negligible. Shear is created at the interface between the fluid and a moving solid surface. Again, because the densitometer vane has very small velocity, we assume that negligible viscous effects are present. Vorticity is generated by both compressibility and viscous shear; so, with both of these neglected, we can assume the flow to be irrotational. The last assumption implies the existence of a velocity potential, and incompressibility implies that this potential satisfies Laplace's equation. This is a major reduction in problem difficulty: Laplace's equation is linear, and it has many methods available for its solution. We consider some quite simple explicit solutions in sections 3.5 and 3.6. Here we use a powerful general result that avoids the necessity for explicit solutions, since it substitutes an overall effect for the interaction of the fluid with the vane. The consequence is a lumped parameter model which has the right properties.

A result from this field of classical hydrodynamics is that, in order to execute nonsteady motions relative to a fluid, a body has to provide two forces: one equal to its own mass times the acceleration, and the second to provide an acceleration to some portion of the fluid surrounding it. This latter term is known as the "added" or "virtual" mass, which is to be added to the body mass in Newton's law of motion (see e.g. [5]). The added mass is found to be proportional to the mass of fluid displaced by the body with a constant of proportionality α, which depends on its shape (α is 1 for a cylinder, $1/2$ for a sphere). This result involving the mass of fluid displaced will bring the density of the fluid into this calculation and will generate a formula relating fluid density and natural frequency.

Consider the free oscillations of the densitometer vane, assuming that it acts as a lumped parameter system with mass M_s and a linear restoring force with spring constant k. The governing equation is

$$(M_s + \alpha M_f)\frac{d^2 y}{dt^2} + ky = 0, \qquad (3.4)$$

where y is a measure of the displacement of the vane from its mean position and M_f is the mass of fluid displaced. Equation (3.4) is the equation
for simple harmonic motion, and the natural frequency ω_n of the system is
given by the square root of $k/(M_s + \alpha M_f)$. Writing $M_f = \rho_f V$, where
ρ_f is the fluid density and V is the volume of fluid displaced, yields, after
rearrangement,

$$\rho_f = \frac{k}{\alpha V \omega_n^2} - \frac{M_s}{\alpha V}. \tag{3.5}$$

This lumped parameter model validates the relation (3.1) between fluid density
and natural frequency, and in a limited sense answers the first of the questions
posed. However, the parameters provided by (3.5) to constitute A and B in
(3.1) are not well defined by the physical parameters of the vane – its dimensions, material properties, etc. Hence this model would need considerable
work to answer the third question in section 3.1, and there do not seem to be
extensions of the model that would answer the second main question posed.
As a consequence a more fundamental modelling approach is attempted: partial differential equations for the motions of the vane and the fluid must be
written down, their interaction examined, and solutions found.

3.4 Fluid–Plate Model

3.4.1 Plate Equation

Consider the plate, length L, breadth B, at rest in the x–z plane, with the
origin located at a corner of the plate attached to the sensing head cylinder,
the z axis along the cylinder-plate join and the x axis pointing along a free
edge of the plate. The transverse displacement of the plate normal to the x–z
plane is denoted by

$$y = w(x, z, t).$$

The partial differential equation governing w is based on

  (i) Linear Theory is applicable: w and its slopes are small so that only
  terms linear in w are kept.

 (ii) The Bernoulli–Euler bending theory of thin plates may be adopted. This
  assumes that longitudinal strains vary linearly across the depth of the
  plate and that the bending moment at a cross-section of the plate is
  proportional to the local radius of curvature. See [6] for details.

The partial differential equation for w is

$$\rho_s h \frac{\partial^2 w}{\partial t^2} + D\nabla^4 w = F \quad \text{in } 0 < x < L, \quad 0 < z < B, \qquad (3.6)$$

where ρ_s is the density of the plate material, h is the plate thickness, and $D = Eh^3/12(1-\sigma^2)$ is the flexural rigidity of the plate, E being the Young's modulus and σ the Poisson ratio of the plate. Numerical values for these constants are given in the Appendix. In (3.6), F is the force applied normal to the plate surface. In our application, F is due to the fluid motion.

3.4.2 Fluid Equation

As in section 3.3, we assume that the fluid is inviscid, incompressible, and irrotational. This implies that the fluid velocity q is the gradient of a potential,

$$q = \nabla\phi,$$

and that ϕ satisfies Laplace's equation

$$\nabla^2\phi = 0. \qquad (3.7)$$

Specification for the domain of applicability of (3.7), which should be restricted in x–y planes over $0 < z < B$ by the cylindrical sensing head, is left to the solution stage, where some strong assumptions are made.

3.4.3 Fluid–Plate Interaction

This is provided by two conditions:

(i) The normal velocity of the plate is identical to the fluid velocity in the y direction at the plate surface. Due to the linearization, the latter may be substituted by the surface $y = 0$. The condition now reads

$$\frac{\partial w}{\partial t} = \frac{\partial \phi}{\partial y} \quad \text{at } y = 0 \qquad (3.8)$$

(ii) The force on the plate, F in (3.6), is due to the difference in fluid pressure above and below the plate, i.e.

$$F = p^- - p^+, \qquad (3.9)$$

where $p^\pm$ denotes the fluid pressure at $y = 0^\pm$. Although the plate has the thickness h in (3.6), h is considered vanishingly small in (3.8) and (3.9), consistent with the linearization.

In order to complete the system, it is necessary to calculate the fluid pressure from other fluid flow quantities. This is provided in classical hydro-dynamics by the Bernoulli equation. Generally this equation contains the square of the fluid velocity and a gravity term, both neglected here, yielding

$$p = -\rho_f \frac{\partial \phi}{\partial t}. \tag{3.10}$$

3.5 Simple Analysis: Incorrect Boundary Conditions

Consider the following (drastic!) assumptions:

(i) The plate has no bending in the z direction and oscillates in the simplest mode in the x direction. The relevant solution to (3.6) is

$$w = b \sin \frac{\pi x}{L} \cos \omega t. \tag{3.11}$$

(ii) The fluid is unconstrained by the cylindrical sensing head housing, so that (3.7) is valid in $y > 0$ and $y < 0$ and the potential ϕ responds to (3.11) through the boundary condition (3.8) interpreted as valid on the whole $y = 0$ plane. Equation (3.8) implies that ϕ has a $\sin(\pi x / L)$ variation; harmonic functions with a sine variation in one direction are exponential in the other. Hence

$$\phi = \begin{cases} \phi^+ = \dfrac{L\omega b}{\pi} \sin \dfrac{\pi x}{L} e^{-\pi y/L} \sin \omega t \text{ in } y > 0, \\[2mm] \phi^- = -\dfrac{L\omega b}{\pi} \sin \dfrac{\pi x}{L} e^{\pi y/L} \sin \omega t \text{ in } y < 0. \end{cases} \tag{3.12}$$

The multiplicative constants in (3.12) have been adjusted to satisfy (3.8), and the exponentials have been chosen to decay as $y \to \pm\infty$. Combining (3.10) and (3.12) allows the pressure above and below the vane to be found, yielding F, the force on the plate, from (3.9). Substituting F and w into (3.6) shows that this equation is satisfied as long as a relation is satisfied between the other parameters in the problem. This condition can be written

$$\rho_f = \frac{D}{2} \left(\frac{\pi}{L}\right)^5 \frac{1}{\omega^2} - \frac{\pi h \rho_s}{2L}. \tag{3.13}$$

Since a solution satisfying both fluid and plate equations, and their interacting boundary conditions, has been found for arbitrary amplitude b, this solution is a homogeneous solution providing "free" oscillations of the system. Hence ω in (3.13) is the system's natural frequency for the mode introduced by (3.11).

(What are other, similar, modes?) The relation in (3.13) between fluid density and natural frequency is the same as (3.1) with

$$A = \frac{D}{2} \left(\frac{\pi}{L}\right)^5, \quad B = \frac{\pi h \rho_s}{2L}.$$

This is an improvement over the lumped parameter model, since now we are able to answer the third main question that was posed. However, we must put (3.13) to the test – how good is it in comparison with data? Table 3.1 shows very disappointing results. (Values of material constants are given in the Appendix.)

The success of this simple model in predicting relation (3.1) indicates that the qualitative features in its formulation seem to be correct, and the poor quantitative predictions encourage us to persist with the main approach of this section, but to examine components of the model to see whether they are reasonable fits to the way the device operates. Note that the mode (3.11) was introduced with no discussion of the boundary conditions to be satisfied at $x = 0, L$ (where the vane is attached to the sensing head). The mode (3.11) was used because it is simple and leads to quick results! Mode (3.11) satisfies

$$w = 0 \text{ and } \frac{\partial^2 w}{\partial x^2} = 0 \text{ at } x = 0, L$$

which are known as *pinned* or *simply-supported* boundary conditions. They imply that the vane has zero displacement and zero moment (or second derivative, i.e. curvature) at both ends. The terminology is descriptive of a plate which has a small hole at each end and is held there by a pin through each hole. This clearly is not accurate for the densitometer vane, which, as described in the Introduction, is brazed into slots of the inner cylinder part of the sensing head. The boundary conditions more appropriate to this situation are known as *clamped*, and the solution for this case is presented in the next section.

Table 3.1. *Frequency Forecasts for the Pinned or Simply Supported Two-Dimensional Model (in Hertz)*

Density (gm/cm^3)	Reported Frequency	Frequency Forecast
0.0	3693	1684
0.7206	2420	940
0.7744	2369	917
0.8114	2322	902
0.8330	2315	893

3.6 Solution with Clamped Boundary Conditions

Clamped boundary conditions assert that the ends of the vane are fixed in the x–z plane:

$$w = 0 \quad \text{and} \quad \frac{\partial w}{\partial x} = 0 \quad \text{at } x = 0, L. \tag{3.14}$$

The mode shape corresponding to these boundary conditions is slightly more complicated than (3.11) – the separation-of-variables solution to the homogeneous part of (3.6) gives a fourth-degree linear ODE for the x variation, and the solutions of this are organized to satisfy (3.14). It may be written

$$w = b \left\{ \cos \beta \left(\frac{x}{L} - \frac{1}{2} \right) \sinh \frac{\beta}{2} + \cosh \beta \left(\frac{x}{L} - \frac{1}{2} \right) \sin \frac{\beta}{2} \right\} \cos \omega t, \tag{3.15}$$

where β satisfies

$$\tan \left(\frac{\beta_i}{2} \right) + \tanh \left(\frac{\beta_i}{2} \right) = 0 \quad i = 1, 2, 3, \dots. \tag{3.16}$$

The three lowest roots of (3.16) corresponding to the first three modes of vibration are $\beta_1 = 4.73$, $\beta_2 = 7.85$, and $\beta_3 = 10.99$.

The solution for the fluid velocity potential is more complicated too – similar to the solution in section 3.5, the harmonic terms in (3.15) give rise to terms exponential in y in the velocity potential, and the exponential terms in (3.15) give rise to harmonic terms in y in ϕ – giving the solution consistent with the boundary condition (3.8):

$$\phi^+(x, t) = \frac{L\omega b}{\beta} \sin \omega t \left\{ e^{-(\beta/L)y} \cos \beta \left(\frac{x}{L} - \frac{1}{2} \right) \sinh \frac{\beta}{2} \right.$$

$$\left. + \left[\cos \frac{\beta}{L} y - \sin \frac{\beta}{L} y \right] \cosh \beta \left(\frac{x}{L} - \frac{1}{2} \right) \sin \frac{\beta}{2} \right\} \quad \text{for } y > 0,$$

$$\phi^-(x, t) = \frac{L\omega b}{\beta} \sin \omega t \left\{ e^{(\beta/L)y} \cos \beta \left(\frac{x}{L} - \frac{1}{2} \right) \sinh \frac{\beta}{2} \right.$$

$$\left. + \left[\cos \frac{\beta}{L} y + \sin \frac{\beta}{L} y \right] \cosh \beta \left(\frac{x}{L} - \frac{1}{2} \right) \sin \frac{\beta}{2} \right\} \quad \text{for } y < 0. \tag{3.17}$$

Again the signs in the exponential terms in (3.17) have been chosen to decay as $y \to \pm\infty$. The harmonic terms in y in (3.17) do not decay as $y \to \pm\infty$; this is a consequence of the solution obtained. The solution for one vane

Table 3.2. *Frequency Forecasts for Clamped–Clamped Two-Dimensional Model (in Hertz)*

Density (gm/cm^3)	Reported Frequency	Frequency Forecast
0.0	3693	3818
0.7206	2420	2431
0.7744	2369	2379
0.8114	2322	2345
0.8330	2315	2326

vibrating would generate velocities decaying at infinity, but an infinite set of vibrating vanes cannot be so restricted. Whether this deficiency in the solution is important is soon seen.

The pressure above and below the vane is now available from (3.10) and (3.17), and F from (3.9), substituted into (3.6), gives that equation satisfied identically when

$$\rho_f = \frac{D}{2}\left(\frac{\beta}{L}\right)^5 \frac{1}{\omega^2} - \frac{\beta h \rho_s}{2L},$$

which is the same as (3.13) with β replacing π. Hence equation (3.1) is again valid with A and B identified. Evaluated at the first mode, $\beta = \beta_1$, table 3.2 shows that the change to clamped boundary conditions has substantially improved the model's ability to predict the relation between fluid density and natural frequency; the fit is almost exact at densities in the region of interest.

The remarks below equation (3.17), worrying about the effect of the solution's non-decay as $y \to \pm\infty$, are now answered by the success in predicting the natural frequency consistent with data. The solution obtained *must* be accurate for the fluid motions in $0 < x < L$ and y small, and the fact that the solution obtained has a nonphysical behavior as either $x \to \pm\infty$ or $y \to \pm\infty$ is unimportant. Presumably a more sophisticated analysis would adopt the solution obtained here as an "inner" solution, and would obtain a suitably matched, decaying solution as an "outer." (See books on matched asymptotic expansion, e.g. [3], for precise explanations of this terminology and procedure.)

3.7 Remarks

Further extensions to include the effects that were posed for consideration by the manufacturer are left to the reader. References [2], [4], and [8] contain a subset of these, including some that were not mentioned earlier.

Some of these are:

- What are three-dimensional effects? The modes considered in sections 3.5 and 3.6 have the vane flat in the z direction; but, clearly, twisting modes are possible. The analysis for such vane modes, with two clamped and two free edges, is not easy, and results must be obtained numerically (see [7]).

- What is the effect of the cylindrical enclosure? Putting the oscillating vane across the diameter of a cylinder and assuming planar motions raises problems for an incompressible fluid, so some of these simplifying assumptions must be relaxed. Which? – and how to retain simple analysis?

- The Clinic team obtained results for the densitometer operating in a gas by incorporating the compressible Bernoulli equation into the model and solving the wave equation instead of Laplace's equation. However the compressible and incompressible frequency forecasts were almost identical, and above the measured values by almost 10%. It would be useful to find a more accurate model for the compressible case. The data are: frequencies 2188, 2138, 2061, 1979 for densities 0.008, 0.023, 0.049, 0.080, respectively.

- The client also mentioned mode contamination (in which the output amplitude being measured at the vane has two peaks very close in frequency) as a cause of wrong identification of the principal natural frequency. Some interesting analysis can be pursued by identifying the frequencies generated by nonlinear terms. Energy is transferred mode to mode by any nonlinearity, and resonance provides amplification at the natural frequencies.

Appendix

The material constants of the vane are:

Length, $L = 3.015$ cm

Width, $B = 1.588$ cm

Density, $\rho_s = 8828$ kg/m^3

Young's modulus, $E = 1.792 \times 10^{11}$ N/m^2

Poisson's ratio, $\sigma = 0.32$

Thickness, $h = 0.071$ cm (for liquid), 0.046 cm (for gas).

References

[1] Borrelli, R. & Coleman, C. (1998) *Differential Equations, A Modeling Perspective*, John Wiley & Sons, New York, USA.

[2] Cumberbatch, E. & Wilks, G. (1987) An Analysis of a Vibrating Element Densitometer, Math. Engng. Ind. **1** 47–66.

[3] Kevorkian, J. & Cole, J. D. (1981) *Perturbations Methods in Applied Mathematics*, Springer-Verlag, New York, USA.

[4] Morris, H., Cumberbatch, E., Al-Dubaibi, A., Everson, R. & Newlin, L. (1984) Analysis of a Vibrating Vane Densitometer. Claremont Graduate School Mathematics Clinic Report, Claremont Graduate University, California, USA.

[5] Paterson, A. R. (1983) *A First Course in Fluid Dynamics*, Cambridge University Press, England.

[6] Southwell, R. V. (1941) *An Introduction to the Theory of Elasticity*, Oxford University Press, England.

[7] Warburton, G. B. (1954) The Vibration of Rectangular Plates, Proc. Inst. Mech. Eng. **168**, 371–384.

[8] Wilks, G., Cumberbatch, E., Burnett, J., Friberg, A., Gassner, S. & Rotenberry, J. (1983) Analysis of a Vibrating Element Densitometer. Claremont Graduate School Mathematics Clinic Report, Claremont Graduate University, California, USA.

Ellis Cumberbatch

Claremont Graduate University, Claremont, CA 91711, USA
ellis.cumberbatch@cgu.edu

4

Acoustic Emission from Damaged FRP-Hoop-Wrapped Cylinders

Preface

In the following case study, a mathematical analysis is made of nonmonotonic behaviour observed in the nondestructive testing of fibre-reinforced pressure vessels by the measurement of acoustic emissions. Although the pressure vessels are subjected to plastic deformation during their manufacture, it was possible to predict nonmonotonic behaviour using linear elasticity for the metal liners and a simple "tension band" model for the fibre wrapping. Not only was the qualitative behaviour predicted, but the mathematical model also gave quantitative results which compared well with experimental results.

Much of the algebraic manipulative work involved in the solution was carried out using the computer algebra software Maple. Familiarity with any one of the several such software packages is highly recommended for any mathematical modeler.

The problem considered was presented by Powertech Labs at the first Study Group with Industry meeting held in Canada under the sponsorship of the Pacific Institute of the Mathematical Sciences. The meeting took place in Vancouver at the University of British Columbia in 1997. Thanks are due to Dr A. Akhtar of Powertech for permission to include material on the testing of the pressure vessels.

4.1 Introduction

Metal cylinders overwrapped by continuous-filament fibre-reinforced plastic (FRP) are used for the storage of compressed gases in aerospace and terrestrial applications requiring lightweight pressure vessels. Examples of terrestrial applications are: breathing-air cylinders used by firefighters and vessels used for the storage of compressed gaseous fuels – in particular, natural gas (CNG) and hydrogen on natural gas vehicles (NGV) and hydrogen vehicles. To maximize the fuel efficiency of the vehicle, the fuel storage cylinder is designed to minimize its mass. Thus at operating pressure it is important that the FRP wrapping carries as much of the load as possible. The minimum design pressure permitted on such vessels can be as low as 2 times the maximum operating pressure, a figure substantially lower than those permitted for example by the ASME pressure vessel code. Consequently such vessels operate in the regime of low cycle fatigue. The dynamic stresses arise from the periodic pressurization and depressurization of the vessel between refuellings, in the case of vehicle fuel storage vessels, and the liner becomes fatigued over a relatively small number of cycles.

The effect of the FRP overwrap on the load distribution remains a significant feature throughout the life cycle of such cylinders. During their fabrication, the vessels are subjected to autofrettage treatment. The FRP-overwrapped vessel is pressurized so as to cause a slight plastic deformation of the metallic liner. When the cylinder is depressurized the liner follows a slightly different stress–strain curve which would leave it with a residual strain in the stress-free state. The wrapping, which has not passed its elastic limit, retraces its original stress–strain path. When the liner and the wrapping attain mutual equilibrium, the load sharing between them results in a residual strain with the liner in compression and the wrapping in tension. When the cylinder is again pressurized, the compression in the liner must be overcome before it carries any tensile load; thus, at service pressure, autofrettage lowers the ratio of the load shared between the liner and the FRP, and the fatigue life of the metallic liner is increased.

Minor flaws developed in the wrapping, for instance by impacts or even in the metal lining, do not render the tank useless but may affect its integrity. Thus a nondestructive means of testing the severity of flaws becomes important, so that tanks which are in danger of rupturing can be identified. Acoustic emission is a possible nondestructive test method which has attractive possibilities for periodic in-situ testing of tanks.

In acoustic emission testing, the cylinder is pressurized to a level about or just above its peak service pressure. Flaws in the wrapping or in the lining lead to a local stress concentration which in turn leads to audible breakage of the reinforcing fibres in that region. The amount of acoustic energy released is recorded. The expectation was that this would reflect the severity of the flaw, the acoustic energy increasing with flaw length, thereby allowing an assessment of the integrity of the vessel. A monotonic relationship between bursting pressure and flaw length had already been established.

Powertech had carried out controlled experiments in which flaws were introduced artificially into the FRP. Cuts were made in the axial direction completely through the FRP. The cuts were of different lengths ranging from

25×10^{-3} m to 250×10^{-3} m, and acoustic emission tests were carried out on the various cylinders. In each case a cylinder was first pressurized to 24.1 MPa, approximately its operating pressure. After depressurization it was repressurized to 27.6 MPa. The procedure used was to increase pressure from 0 to 27.6 over a period of 5–7 mins, and then keep the pressure steady for 5 mins before depressurizing. Results for two cylinders are shown in figures 4.1 and 4.2. In each case the acoustic energy measured in the pressurization phase is denoted by the square, and the amount in the holding phase by the diamond. The cumulative amount is shown by the circle. The data points are joined (perhaps unwisely) using cubic splines. Results for a steel-lined cylinder are given in figure 4.1, and for an aluminum alloy liner in figure 4.2.

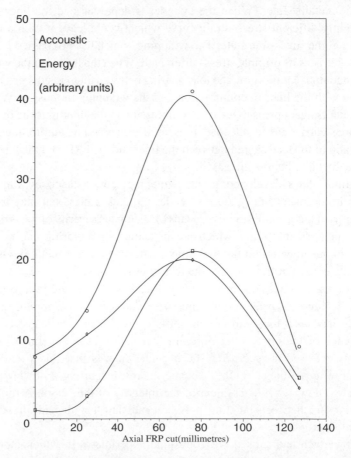

Fig. 4.1. Experimental acoustic emissions for a steel lined cylinder: acoustic energy vs. flaw length

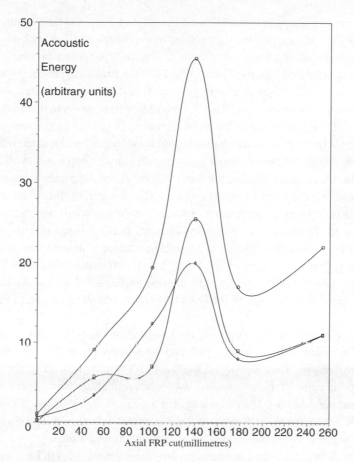

Fig. 4.2. Experimental acoustic emissions for an aluminum-lined cylinder: acoustic energy vs. flaw length

As can be seen from figures 4.1 and 4.2, the acoustic emission energy did increase with the length of the flaw at first, but then a maximum was reached at a critical length, with a subsequent decrease in the energy of the acoustic emissions as the flaw length increased further.

The problem proposed by Powertech was to explain this nonmonotonic behaviour of the energy with the flaw length.

4.2 Problem Description

The FRP wrapping consists of many parallel fibres embedded in a comparatively weak plastic matrix, and provides little resistance to stretching in the direction perpendicular to the fibres. For the cylinders under consideration, the fibres are wrapped in the circumferential direction, and bulk of the forces in the FRP are carried by the tension in the fibres and are entirely in the circumferential or "hoop" direction. When the cylinder is pressurized, its radius is increased, and the fibres in the FRP are stretched, producing tension in the fibres. The liner is also under stress, and the pressure load is shared between the wrapping and the hoop (circumferential) stress in the liner. Since the "artificial flaws" are in the axial direction, the effect on the wrapping is to release the tension in a circumferential band, whose width is the length of the cut. When the tank is under pressure, this unwrapped region will bulge, and the fibres at the edge of the band will be under the greatest tension and therefore be the most likely to rupture. The number of fibres that rupture should be related to the tension at the edges of the band, which in turn is related to the circumferential displacement. It is expected that the acoustic energy released is related, at least monotonically, to the circumferential displacement at the edge of the band. The nonmonotonic behaviour of the energy with the flaw length could then be explained if the circumferential displacement at the edge of the band is a nonmonotonic function of the flaw length.

Since there is no tension along the length of the flaw, this part of the liner may be considered to be unwrapped, and the wrapping here plays no part in the physical model. The resulting problem then is to examine the displacement in a metallic cylinder under pressure, with reinforcing fibre wrapping except for a circumferential band of given length at the location of the flaw. To simplify the model, autofrettage is ignored, so that, under a condition of no strain, the tension in the wrapping and the stresses in the liner are zero.

The cylinders under consideration are approximately 1.6 m long and have radii of about 0.15 m. The thickness of the metal liners is 0.0063 m for a steel liner and 0.013 m for an aluminum alloy liner. The flaw lengths range from 0.025 m to 0.250 m. For simplicity it is assumed that the metallic cylinder is linearly elastic. For the modeling we make the following observations.

- Since the thickness to radius ratio is small (.04 to .08), elastic shell theory may be used.
- Since the flaw length to cylinder length is small (about 1/6), the cylinder is considered to be infinite.
- The internal pressure is axisymmetric, as is the pressure on the liner due to the FRP. Thus the problem is axisymmetric, and all variables are functions of the axial coordinate x only.
- There will also be symmetry about the centre of the unwrapped band, and this is taken as the origin $x = 0$ of the x coordinate (see figure 4.3).

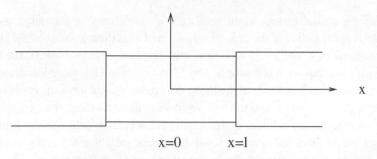

Fig. 4.3. The cylinder with the x coordinate system

The equations for linearly elastic cylindrical shells with axisymmetry can be found in Timoshenko & Woinowsky-Krieger [3].

They are the equilibrium equations

$$\frac{d}{dx} N_x = 0 \tag{4.1}$$

$$\frac{d}{dx} Q_x + \frac{N_\phi}{a} + Z = 0 \tag{4.2}$$

$$\frac{dM_x}{dx} \quad Q_x = 0 \tag{4.3}$$

and the stress displacement relations

$$N_x = \frac{E_s h_s}{1 - v^2} \left[\frac{\partial u}{\partial x} - v\frac{w}{a} \right] \tag{4.4}$$

$$N_\phi = \frac{E_s h_s}{1 - v^2} \left[-\frac{w}{a} + v\frac{\partial u}{\partial x} \right] \tag{4.5}$$

$$M_x = \frac{-E_s h_s^3}{12(1 - v^2)} \frac{\partial^2 w}{\partial x^2}. \tag{4.6}$$

The displacements at the mid surface of the shell are $u(x)$ in the axial direction and $w(x)$ in the inward radial direction. E_s is the Young's modulus and v the Poisson's ratio for the shell. The radius of the shell is a and the thickness is h_s. Z is the force per unit area acting on the shell in the inward normal (radial) direction. Z is comprised of the internal pressure $-p_0$ and the inward normal pressure on the liner due to the FRP wrapping.

To obtain the equilibrium equations, the three-dimensional equations for the axisymmetric equilibrium of a shell in cylindrical polar coordinates are integrated from $-h_s/2$ to $h_s/2$ with respect to the inward radial distance z

from the central surface of the shell ($z = a - r$, where r is the radial coordinate, a is the radius of the central surface, and h_s is the thickness of the shell). Equations (4.1) and (4.2) are the result. To obtain equation (4.3), the axial equilibrium equation is multiplied by z, and the result integrated with respect to z from $-h_s/2$ to $h_s/2$. In the three-dimensional equations, the coordinates are $x, \phi, z\, (= a - r)$, and the unit vectors are directed along the axis and circumference of the cylinder and in the inward normal direction to the cylinder. Because of the axial symmetry, all dependent variables are independent of ϕ, and the circumferential displacement is zero. Of the stresses τ_{ij}, the shear stresses $\tau_{x\phi}$ and $\tau_{\phi z}$ are zero.

In the equations (4.1–4.3), N_x, N_ϕ, Q_x are stress resultants given by

$$N_x = \int_{-h/2}^{h/2} \tau_{xx}\, dz, \qquad N_\phi = \int_{-h/2}^{h/2} \tau_{\phi\phi}\, dz, \qquad Q_x = \int_{-h/2}^{h/2} \tau_{xz}\, dz$$

and M_x is the moment resultant given by

$$M_x = \int_{-h/2}^{h/2} z\tau_{xx}\, dz.$$

To obtain the stress displacement relations (4.4–4.6), the standard shell assumptions that τ_{xz} and τ_{zz} are small compared to the other stresses are used. This leads to the axial displacement being linear in z and given by

$$u(x) = z\frac{\partial w(x)}{\partial x}$$

and the strain

$$\epsilon_{zz} = -\nu(\epsilon_{xx} + \epsilon_{\phi\phi})/(1 - \nu).$$

When these assumptions are used with the three-dimensional stress displacement, (4.4–4.6) relations result.

Alternative derivations of the equations for elastic shells using asymptotic analysis can be found in Green [2] and Goldenweiser [1].

Equation (4.1) immediately gives $N_x = $ constant. We expect that the constant value of N_x is due to tension produced by the action of the internal pressure p_0 on the ends of the tank. Since the pressure on the curved end surfaces is constant and is in the direction normal to the surface, the resultant force on an end can be calculated by the use of Green's theorem. The force is $\pi a^2 p_0$ in the outward axial direction. This force must be balanced by the total normal stress resultant $2\pi a N_x$. Thus

$$N_x = \frac{a p_0}{2}.$$

Using this in equations (4.4) and (4.5), we see that

$$N_\phi = -E_s h_s \frac{w}{a} + \frac{vap_0}{2}, \quad \text{and} \quad \frac{du}{dx} = v\frac{w}{a} + \frac{ap_0(1 - v^2)}{2E_s h_s}.$$

Further using (4.4) and (4.6), we see that

$$Q_x = \frac{-E_s h_s^3}{12(1 - v^2)} \frac{d^3 w}{dx^3}.$$

Finally substituting all of these results into (4.2), we obtain

$$\frac{-E_s h_s^3}{12(1 - v^2)} \frac{d^4 w}{dx^2} - E_s h_s \frac{w}{a^2} + \frac{vp_0}{2} + Z = 0.$$

We now examine the contribution of the FRP wrapping to Z. We suppose that there is no wrapping for $-l \le x \le l$ and thus, in this interval, $Z = -p_0$. Outside this interval the FRP will make a contribution to Z which we now consider.

We let T be the resultant force per unit length along the cylinder due to a band of fibre of thickness h_f (see figure 4.4). The resulting inward normal pressure p on the liner, which is constant in the circumferential direction, satisfies the force balance equation

$$\int_{-\theta}^{\theta} (p \cos t) a dt = 2T \sin \theta.$$

This gives

$$2pa \sin \theta = 2T \sin \theta \quad \text{or} \quad p = \frac{T}{a}.$$

The strain in a circumferential band is given by $-2\pi w/(2\pi a)$, and thus T is given by $T = -E_f h_f w/a$, where E_f is the Young's modulus of the reinforcing fibre. Because we have ignored the autofrettage, we assume no pre-tension in the wrapping nor initial compression in the liner. Thus for $|x| > l$, we have $Z = -p_0 - E_f h_f w/a^2$, and our final equation for w is given by

$$\frac{-E_s h_s^3}{12(1 - v^2)} \frac{d^4 w}{dx^4} - E_s h_s \frac{w}{a^2} - (1 - \frac{v}{2})p_0 = 0 \qquad |x| \le l \qquad (4.7)$$

$$\frac{-E_s h_s^3}{12(1 - v^2)} \frac{d^4 w}{dx^4} - (E_s h_s + E_f h_f)\frac{w}{a^2} - (1 - \frac{v}{2})p_0 = 0 \qquad |x| > l.$$

$$(4.8)$$

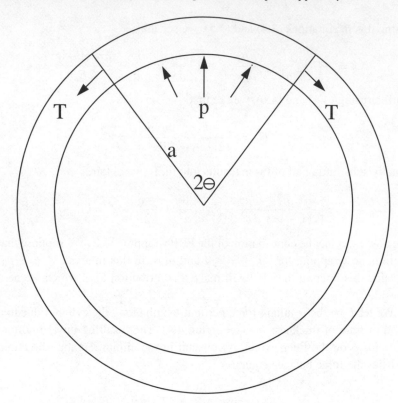

Fig. 4.4. Pressure on the liner due to tension in the FRP

Because of symmetry about $x = 0$, we consider only the region $0 \le x < \infty$ and have the boundary conditions

$$\frac{dw}{dx} = 0, \frac{d^3w}{dx^3} = 0 \quad \text{at} \quad x = 0.$$

As "$x \to \infty$" we approach the ends of the cylinder. We do not expect the boundary conditions at the ends to affect the solution close to the origin, so we impose the condition that $w \to$ constant exponentially. The coefficient of w/a^2 is discontinuous at $x = l$, so we must also impose continuity on

$$w, \frac{dw}{dx}, M_x \quad \text{and} \quad Q_x \quad \text{at} \quad x = l.$$

Because M_x is a constant multiple of d^2w/dx^2 and Q_x is a constant multiple of d^3w/dx^3, this is equivalent to requiring that w and its first three derivatives be continuous at $x = l$. (This condition is often written as $w \in C^3$ at $x = l$.)

4.3 Problem Solution

We now nondimensionalize the differential equations (4.7) and (4.8). We let

$$W = \frac{-wE_s h_s}{\left(1 - \frac{v}{2}\right) p_0 a^2}, \quad \lambda = \frac{E_s h_s + E_f h_f}{E_s h_s}$$

and $\xi = x/b$ where

$$b = \left(\frac{h_s^2 a^2}{3(1 - v^2)}\right)^{1/4}.$$

The scaling of w (the sign has been changed so that W is positive when the cylinder expands) and the choice of λ are reasonably obvious. The choice of a length scale proportional to $\sqrt{ah_s}$ is common in shell theory and allows a balance between the w and d^4w/dx^4 terms. Using a prime to denote differentiation with respect to ξ, the equations become

$$\frac{1}{4}W'''' + W = 1 \quad (0 \le \xi \le L)$$

$$\frac{1}{4}W'''' + \lambda W = 1 \quad (L < \xi) \quad (4.9)$$

where $L = l/b$. The boundary conditions become:

$$W'(0) = W'''(0) = 0, \quad W \to \text{constant as } \xi \to \infty$$

$$W, W', W'', W''' \text{ continuous at } \xi = L.$$

(The factor of $1/4$ appearing in the equations is a mathematical convenience.)

The general solution of equation (4.9) is, with $\beta = \lambda^{\frac{1}{4}}$,

$$W = \frac{1}{\beta^4} + e^{-\beta\xi}(A\cos\beta\xi + B\sin\beta\xi)$$

$$+ e^{\beta\xi}(C\cos\beta\xi + D\sin\beta\xi) \quad \xi \ge L.$$

The condition at infinity requires that $C = D = 0$, so that these terms do not affect the solution near $\xi = 0$.

For convenience in requiring continuity at $\xi = L$, we rewrite the solution for $\xi \ge L$ as

$$W = \frac{1}{\beta^4} + e^{-\beta(\xi - L)}(a\cos\beta(\xi - L) + b\sin\beta(\xi - L)).$$

The general solution in $0 \le \xi \le L$ is written in terms of hyperbolic cosine and sine, and a convenient form which takes account of the evenness in ξ is

$$W = 1 + A\cosh\xi\cos\xi + B\sinh\xi\sin\xi \quad 0 \le \xi \le L.$$

Imposition of the continuity conditions at $\xi = L$ leads to the following system of equations for A, B, a, b.

$$
\begin{bmatrix}
Cc & Ss & -1 & 0 \\
Sc - Cs & Sc + Cs & \beta & -\beta \\
-2Ss & 2Cc & 0 & 2\beta^2 \\
-2(Sc + Cs) & 2(Sc - Cs) & -2\beta^3 & -2\beta^3
\end{bmatrix}
\begin{bmatrix}
A \\ B \\ a \\ b
\end{bmatrix}
=
\begin{bmatrix}
\dfrac{1}{\beta^4} - 1 \\ 0 \\ 0 \\ 0
\end{bmatrix}
$$

(4.10)

where $C = \cosh L$, $S = \sinh L$, $c = \cos L$, $s = \sin L$. To examine the nonmonotonic behaviour, we must obtain $W(L)$ as a function of L. Thus $W(L)$ is given by $W(L) = a + 1/\beta^4$, where a must be found from the equations (4.10).

It is possible to obtain asymptotic solutions of equations (4.10) in powers of L for small L and in powers of e^{-L} for large L. This is left as an exercise for the reader.

To carry out the substantial amount of algebraic manipulation required to solve the equations (4.10) exactly, Maple was invoked, and the result was

$$
W(L) = \frac{(\beta^4 + 1)\Gamma_- + 2\beta^3(C^2 - c^2) + 2\beta^2\Gamma_+ + 2\beta(S^2 + c^2)}{\beta^2[(\beta^4 + 1)\Gamma_+ + 2\beta^3(S^2 + c^2) + 2\beta^2\Gamma_- + 2\beta(C^2 - c^2)]}
$$

where

$$
\Gamma_+ = CS + cs, \quad \Gamma_- = CS - cs.
$$

To display the nonmonotonic nature of the function $W(L)$, Maple was again used to plot $W(L)$ against L. The results for different values of λ (or β) are shown in figure 4.5. Nonmonotonic behaviour is seen for all values of λ.

The model was compared with two sets of experimental results supplied by Powertech (figures 4.1, 4.2).

The first was for a steel-lined tank with $E_s = 203{,}000$ MPa, $E_f = 43{,}700$ MPa, $h_s = 6.3 \times 10^{-3}$ m, $h_f = 7.8 \times 10^{-3}$ m, $a = 0.1635$ m, and $\nu = 0.28$. This gives $\lambda = 1.266$ and $b = 0.02489$ m. The maximum displacement $W(L)$ occurs at $L = 1.207$ for a crack length of $2bL = 0.06008$ m. This is in good correspondence with the experimental result of figure 4.1.

The second was for an aluminum alloy tank with $E_s = 71{,}700$ MPa, $E_f = 43{,}700$ MPa, $h_s = 13.8 \times 10^{-3}$ m, $h_f = 6.3 \times 10^{-3}$ m, $a = 0.152$ m and $\nu = 0.30$. For this tank, $\lambda = 1.278$ and $b = 0.03563$ m.

The maximum value of $W(L)$ occurs at $L = 1.209$ for a crack length of $2bL = 0.086$ m. This is a little too small when compared to the experimental result of figure 4.2.

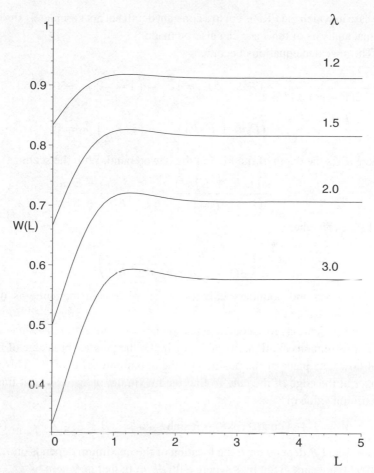

Fig. 4.5. Displacement $W(L)$ at the edge of the crack as a function of crack length L for different values of λ

Examining the graphs of $W(L)$ for different values of λ, it is seen that the value of L which gives the maximum value of $W(L)$ changes very little, and it would seem that a value of L between 1.2 and 1.25. would be appropriate for most cases. This means that a crack length of $2.4b$ to $2.5b$ would give the maximum acoustic energy for all cases. The length b depends only on h_s, a, and ν and not on h_f, E_f, or E_s.

4.4 Further Analysis

In the analysis so far, it has been assumed that the FRP wrapping is cut completely through. The scaled equations, however, remain unchanged for the case in which the FRP is cut to a constant depth but not completely through, so that analysis of this case can also be made.

The unscaled equations become

$$\frac{-E_s h_s^3}{12(1-v^2)}\frac{d^4 w}{dx^4} - (E_s h_s + E_f h^*)\frac{w}{a^2} - \left(1 - \frac{v}{2}\right)p_0 = 0 \qquad |x| \le l$$

$$\frac{-E_s h_s^3}{12(1-v^2)}\frac{d^4 w}{dx^4} - (E_s h_s + E_f h_f)\frac{w}{a^2} - \left(1 - \frac{v}{2}\right)p_0 = 0 \qquad |x| > l.$$

where h^* is the depth of the FRP in the flawed band. With the scalings

$$W = \frac{-w(E_s h_s + E_f h^*)}{(1 - \frac{v}{2})p_0 a^2}, \quad \lambda = \frac{E_s h_s + E_f h_f}{E_s h_s + E_f h^*}$$

and $\xi = x/b$ where

$$b = \left(\frac{h_s^2 a^2 E_s h_s}{3(1-v^2)(E_s h_s + E_f h^*)}\right)^{1/4}$$

the equations and boundary conditions for W remain the same as in the previous section.

The implications for acoustic emission however are different. Since some FRP now remains at all points of the cylinder, the greatest breakage of fibres will now take place at the position of the maximum value of W. This is no longer at the edge of the cut. To find the maximum value, we must find the maximum value of

$$W = 1 + A \cosh \xi \cos \xi + B \sinh \xi \sin \xi \qquad 0 \le \xi \le L. \tag{4.11}$$

Since A and B depend on L the location of the maximum depends on L. The critical points are given by ξ where $dW/d\xi = 0$; that is, where

$$(A + B) \sinh \xi \cos \xi - (A - B) \cosh \xi \sin \xi = 0 \qquad (0 \le \xi \le L).$$

Clearly $\xi = 0$ is a solution for all L. If $A - B = 0$, then $\xi = \pi/2, 3\pi/2, \dots$ are solutions. If $A + B = 0$, then $\xi = \pi, 2\pi, \dots$ are solutions. For other values of A and B, critical points ξ are solutions of

$$(A + B) \tanh \xi = (A - B) \tan \xi. \tag{4.12}$$

The location and nature of the critical points, the solutions of (4.12), depends on the signs and relative magnitudes of $A + B$ and $A - B$. Graphs of $A + B$ and $A - B$ as functions of L are given in figure 4.6 for $\lambda = 1.2$. It is found that, for L small, $A < 0$, $B < 0$, and $A + B < A - B < 0$. hence there is only one maximum value of W in the interval $[0, L]$, and it is at $\xi = 0$, where $W = 1 + A$. When L increases beyond the first value

at which $A + B = A - B$ or $B = 0$, a new critical point appears inside
the interval $[0, L]$ as a solution of equation (4.12). From figure 4.6 it is seen
that $A + B > A - B$ after the first point where $B = 0$. An examination
of the intersection of the curves $(A + B) \tanh \xi$ and $(A - B) \tan \xi$ shows
that, after this value of L, the new solution gives a local maximum and
$\xi = 0$ becomes a local minimum. As L increases further, it is found that
the value of ξ at which the maximum occurs moves away from the origin.
Eventually, as L reaches the next value at which $B = 0$, the origin again
becomes a local maximum but not the global maximum. This is shown in
figure 4.7, where the solution $W(\xi)$ of (4.11) is plotted as a function of ξ for
$L = 1, \dots, 8$.

The maximum value of W can then be found as a function of L. To achieve
this, A and B must be evaluated for the given value of L, and then the critical
point is found by solving equation (4.12). Finally the value of W at the critical
point is found. This must be repeated for all L. This arithmetically unpleasant

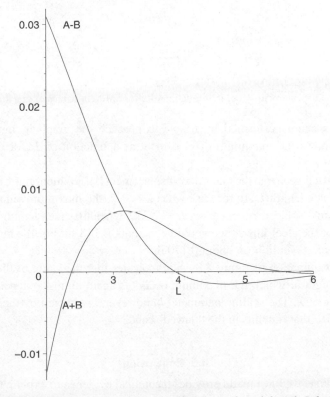

Fig. 4.6. The quantities $A + B$ and $A - B$ as functions of crack length L for $\lambda = 1.2$

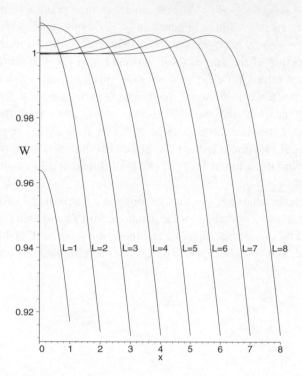

Fig. 4.7. Displacement $W(x)$ as a function of x for different values of crack length L

task was again performed by Maple to produce the graph in figure 4.8, which shows the maximum displacement as a function of L for different values of λ.

Again it is seen that the maximum displacement is a nonmonotonic function of the crack length L. In the case when $\lambda = 1.2$, the maximum value of the maximum displacement occurs at $L = 2.379$, when the crack length is about $4.8b$. For the steel liner, this would be at about 0.118 m. For the aluminum alloy liner, it would be at about 0.170 m.

An examination of figure 4.8 again indicates that the value of the scaled crack length at which the maximum occurs is again almost independent of the value of λ. The scaling parameter b, however, is affected by the amount of the FRP that remains in the "flawed" band.

4.5 Conclusion

The analysis we have made provides theoretical evidence to explain the non-monotonic behaviour observed in the experiments. It suggests that a length

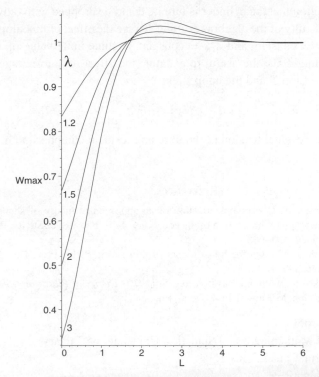

Fig. 4.8. Maximum displacement W_{max} as a function of crack length L for different values of λ

scale b proportional to $\sqrt{ah_s}$ is a major factor in the behaviour of the tank. For crack lengths less than $2b$, the displacement at the crack edge increases with crack length until a maximum is reached. For crack lengths greater than $4b$, the displacement breaks down into three segments: a central section where the displacement is essentially that of a cylinder made of the liner only, a transition region of length of order b which contains the maximum bulge and the crack edge, and a section where the displacement is essentially that of the undamaged wrapped cylinder.

The simplicity of the model and our ability to obtain analytical solutions is due to the assumption of axisymmetry and linear elasticity. Cuts of varying depth or an assumption of nonlinear elastic behaviour would require that the ordinary differential equations would have to be solved numerically. Non-axisymmetric loading or nonaxisymmetric fibre wrapping (which does occur in practice) would require a two-dimensional shell theory and the numerical solution of the resulting partial differential equations.

The assumption that the cylinder is infinite could be dropped with only an increase in difficulty of the algebra resulting. To see the effect of this, impose conditions $Q_x = $ constant and $M_x = $ constant, at some finite value of x of order 1 m. It might also be useful to examine the effect of autofrettage by changing the tension T and the hoop stress N_ϕ to

$$T = -E_f h_f \frac{w}{a} + T_0, \qquad N_\phi = -E_s h_s \frac{w}{a} + \frac{\nu p_0}{2} - T_0$$

where T_0 is the residual tension in the fibre and residual compression in the liner.

References

[1] Goldenweiser, A. L. (1963) Derivation of an approximate theory of shells by means of asymptotic integration of the equations of the theory of elasticity. Prikl. Mat. Mekh. **27**, 593–608.

[2] Green, A. E. (1962) On the linear theory of thin elastic shells. Proc. Roy. Soc. Lond. A **226**, 143–160.

[3] Timoshenko, S. & Woinowsky-Krieger, S. (1970) *Theory of Plates and Shells*, Second Edition, McGraw-Hill, London, England.

D. Rex Westbrook

Department of Mathematics and Statistics, University of Calgary, Calgary, Alberta, Canada T2N 1N4

westbroo@ucalgary.ca

5

Modelling the Cooking of

a Single Cereal Grain

Preface

The following case study develops several models to help understand how to uniformly and accurately cook whole grains for the manufacture of breakfast cereals. During the cooking process, heat and moisture must enter the grain. These two transport processes are modelled by diffusion equations. Linear diffusion is appropriate for the heat transport. A nonlinear diffusion model is examined for the more complex process of water uptake in the grain. The times to heat and wet the grain can be estimated using both numerical and analytic approaches. The mean action time proves to be a very powerful way of estimating the speed of the wetting front. The degree of overcooking in the present manufacturing process can also be estimated. Finally, some recommendations for improving the cooking process are suggested.

The work outlined here was part of the analysis carried out at the 1996 Mathematics-in-Industry Study Group at the University of Melbourne, Australia. Some extensions of the work have been carried out by various groups ([3], [5]). Only the basic analysis is presented here; extensions are suggested as projects. The collaboration with the company is ongoing.

5.1 Introduction

Starch crops, such as cereals, pulses, and tubers, account for over half the total food produced and consumed in the world. Humans find it easiest to digest starch after it has been cooked. The application of both heat and moisture to starch causes it to gelatinise – this allows the starch to be able to be digested by enzymes.

The cooking of intact whole grains (e.g. wheat, corn, rice) is industrially important in breakfast cereal manufacture. Australians are the world's largest consumers per capita of ready-to-eat breakfast cereals, consuming more than Aus\$ 650 million worth each year. The cooking of single whole grains is the first stage in the manufacture of breakfast cereal.

Uncle Tobys is an Australian-owned food manufacturing company which has existed for more than 100 years. At present Uncle Tobys claims about a quarter of the Australian breakfast cereal market, and this makes up about half its business.

The typical breakfast cereal manufacturing process is outlined here. Whole cereal grains of corn, wheat, oats, and rice are cooked at the Uncle Tobys Wahgunyah Plant (in northeastern Victoria) in a rotating steam pressure cooker, in one-tonne batches. Cooking depends on two concurrent processes: heating the grain to a high enough temperature, and hydrating the grain to a high enough moisture content. After cooking, the grains are dried, flaked, toasted, and coated before being packaged as a finished breakfast cereal product.

At present, the manufacturing line for breakfast cereal at the Wahgunyah plant is operating at near full capacity. The rotary cooking operation is known to be the bottleneck in the line capacity. By understanding the cooking process, it is hoped that strategies to increase the capacity of the line can be developed.

Besides speeding up the cooking process, Uncle Tobys stressed the importance of being able to ensure a uniformly cooked product. Under-cooking of part of the grain produces a raw taste and poor binding; and overcooking produces a sticky dough that is difficult to process, and a product that goes soggy too quickly in milk.

5.2 The Problem

The Uncle Tobys wanted to determine how to cook grains for the best texture and taste. The company was looking to speed up the cooking process and ensure its grains were evenly cooked. Two transport processes occur during the cooking – both heat and moisture must enter the single grain.

Uncle Tobys came with the following set of questions and goals:

- What is the temperature within a single cereal grain as a function of the following variables: grain composition, grain geometry, process temperature, process pressure, and time?
- What is the moisture distribution within a single cereal grain as a function of the above variables?
- Is the heat transfer a slower process than the moisture uptake?
- Can the degree of overcooking in a single grain be deduced as a function of the above variables?

• Can recommendations for improving and speeding up the cooking process be made: for example, would it help to separate the two transport processes?

5.3 Background

In order to understand the cereal cooking process, it is necessary to have some knowledge of the structure of a typical cereal grain. All grains are covered in an outer bran layer called a pericarp, which has relatively low water permeability. The majority of the grain interior is the starch endosperm, which consists of bundles of starch granules surrounded by a protein matrix. Figure 5.1 gives a schematic representation.

Gelatinisation is the process whereby water and heat cause starch granules to swell, and eventually burst at the gelatinisation temperature T_{gel}. The exact mechanism of gelatinisation continues to be the subject of debate. Excellent reviews of the process may be found in [8] and [10].

If a starch granule is heated in an aqueous environment, initially the granule wall is able to swell to accommodate the water. The swelling is reversible up to a temperature somewhat below the bursting temperature, T_{gel}, and irreversible beyond it when the elastic limit is exceeded. This is the beginning of gelatinisation. At some point the granule bursts, releasing a viscous starch paste. This released starch is a very delicate material. The long chains of the starch are easily broken into shorter units if the paste is subject to excessive mechanical or thermal treatments. This is detrimental to starch quality (and affects the crispness of the resulting product). Thermal degradation begins as soon as the granule bursts, and continues as long as the starch is at an elevated temperature. Starch damage leads to unpleasant (and therefore unpopular) breakfast cereal flake textures.

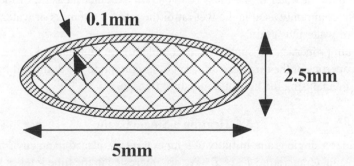

Fig. 5.1. Cross-section of a simplified rice grain

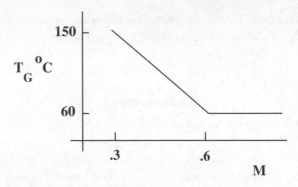

Fig. 5.2. A sketch indicating gelatinisation temperature versus moisture content (dry ratio)

Table 5.1. *Some Properties of Whole Grains*

Properties	Corn Cellulose	Corn Starch	Units
specific heat C_p	2500	2160	J/(kg K)
density ρ	950	1400	kg/m^3
thermal conductivity λ	0.15	0.26	W/(m K)
thermal diffusivity α	6.3×10^{-8}	8.6×10^{-8}	m^2/s
moisture diffusivity D at 40°C [9]	$6 \times 10^{-12} \, e^{8.6M}$	$1.5 \times 10^{-11} \, e^{8.6M}$	m^2/s
length scale	0.1–0.15	1.25	mm

Gelatinisation occurs over a temperature range of about 10°C, at a temperature that depends on the moisture level. The relationship is illustrated in figure 5.2. If moisture content is too low, it is thought that gelatinisation cannot occur at all. The higher the moisture content, the lower the temperature at which gelatinisation occurs. Moisture content is here the dry ratio: the ratio of the mass of water to the mass of dry grain. Another measure of moisture content commonly used is the wet ratio: the ratio of the mass of water to the mass of water plus grain.

Grain properties are summarised in table 5.1. In this analysis, we assume the properties of the thin pericarp are not important, to the dominant order at least, in analysing the heat and moisture transport.

5.4 Heating a Single Grain

Consider a single grain, initially at temperature T_0, placed in an environment at constant temperature $T_1 > T_0$. We are interested in the time it takes for the grain to heat up.

The linear diffusion equation for temperature $T(\mathbf{r}, t)$ at time t and at location $\mathbf{r}$ in the grain is

$$\frac{\partial T}{\partial t} = \alpha \nabla^2 T \qquad (5.1)$$

with initial temperature $T(\mathbf{r}, 0) = T_0$, and $T = T_1$ at the grain surface for $t > 0$. This has solutions that depend on the grain shape.

Two different measures of time to heat are used here. These time measures are illustrated in figure 5.3. Since the grains we consider have spherical or ellipsoidal shapes, the centre of the grain will be the slowest point to heat up. Let t^* be the time for the temperature at the centre of the grain to move to 90% of its final value. We also introduce the mean action time, t^{**}, which is particularly useful for the nonlinear wetting problem in later sections ([6], [7]). A particular advantage of this approach is that it yields the solution's dependence on the parameters of the problem. This is very difficult for non-linear problems, but the use of mean action time leads to a linear problem which can be solved.

A solid theoretical basis for heating time is given by McNabb and McNabb and Wake ([6], [7]), and is outlined briefly here. For a linear diffusion process, the mean action time t^{**} is defined at each point in the grain by

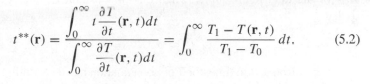

$$t^{**}(\mathbf{r}) = \frac{\displaystyle\int_0^\infty t \frac{\partial T}{\partial t}(\mathbf{r}, t)\,dt}{\displaystyle\int_0^\infty \frac{\partial T}{\partial t}(\mathbf{r}, t)\,dt} = \int_0^\infty \frac{T_1 - T(\mathbf{r}, t)}{T_1 - T_0}\,dt. \qquad (5.2)$$

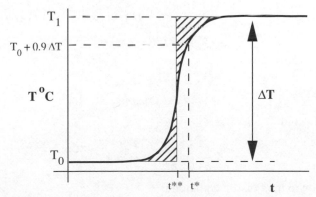

Fig. 5.3. Sketch showing the temperature profile with time, at the centre of the grain. The locations of the time t^* to reach 90% of the final temperature, and of the mean action time t^{**}, are indicated

We define $\Phi(\mathbf{r})$ by setting $t^{**}(\mathbf{r}) \equiv \Phi(\mathbf{r})/\alpha$. Then substitution into the diffusion equation gives the result that Φ satisfies a Poisson equation,

$$\nabla^2 \Phi = -1 \tag{5.3}$$

inside the grain, with $\Phi = 0$ on the grain surface. The largest value of Φ, denoted Φ_G, gives the mean action time for heat to penetrate to the centre of the grain. A graphical interpretation is given in figure 5.3. The areas shaded on either side of the mean action time are equal, showing how it corresponds to locating a front in shock theory. With a sharp front, as occurs in nonlinear diffusion problems like the wetting problem introduced later in this report, there is very little difference between t^* and t^{**}. Henceforth we use t^{**} to mean $t^{**}(\mathbf{r})$ evaluated at the centre of the grain, $\mathbf{r} = 0$. With a more diffuse front, like the present linear problem, t^* can be up to twice as large as t^{**}.

5.4.1 Sphere

For a sphere of radius R, the heat conduction problem is a standard one; see for example [1]. The curves plotted in figure 5.4 show the dimensionless temperature $(T - T_0)/(T_1 - T_0)$ in a sphere versus dimensionless radius r/R, at different dimensionless times $\alpha t/R^2$. The centre of the grain reaches 90% of the final temperature, when this dimensionless time is about 0.3, so that

$$t^* = 0.3 R^2/\alpha. \tag{5.4}$$

For the mean action time, the Poisson equation has the solution

$$\Phi = (R^2 - r^2)/6, \tag{5.5}$$

$$t^{**} = R^2/(6\alpha). \tag{5.6}$$

5.4.2 Ellipsoid

For an ellipsoid

$$\frac{x^2}{a^2} + \frac{y^2}{b^2} + \frac{z^2}{c^2} = 1 \tag{5.7}$$

the solution to the Poisson equation is

$$\Phi = \left[1 - \left(\frac{x^2}{a^2} + \frac{y^2}{b^2} + \frac{z^2}{c^2} \right) \right] / A, \tag{5.8}$$

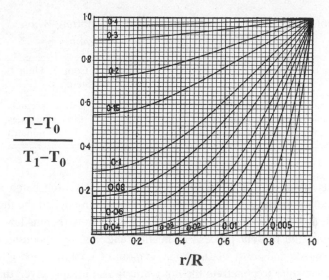

$$\frac{T-T_0}{T_1-T_0}$$

r/R

Fig. 5.4. Temperature profiles at different dimensionless times $\alpha t / R^2$ for heat conduction into a sphere of radius R

where

$$A = 2\left(\frac{1}{a^2} + \frac{1}{b^2} + \frac{1}{c^2}\right). \tag{5.9}$$

The penetration of heat then follows ellipsoids, successively smaller and smaller as time progresses, like onion skins. The mean action time for heat to penetrate to the centre of an ellipsoidal grain is

$$t^{**} = \frac{1}{2\alpha\left((1/a^2) + (1/b^2) + (1/c^2)\right)}, \tag{5.10}$$

and t^* is approximately twice this.

5.5 Timescales for Wetting and Heating – Linear Models

The wetting of grain starch is a nonlinear process, and is dealt with as such in the following section. Here we obtain quick estimates to compare the times to heat and to wet, by approximating grain wetting as a linear diffusion process with constant diffusion coefficient $D = D_0 \exp(8.6M)$ with (a) $M = 0.15$ and (b) $M = 0.65$.

The approximate times t^* to heat and wet grains calculated in this way are presented in table 5.2. These provide only a first approximation to the times involved in wetting a grain. However, it is immediately clear that a grain heats up in a much shorter time than it takes to get wet.

Table 5.2. *Some Timescales for Heating and Wetting Whole Grains, Using the Linear Approximation*

Times t^*	Cellulose	Corn Starch	Units
timescale – linear heating	0.05	5.4	seconds
timescale – linear wetting ($M = 0.15$)	2	145	minutes
timescale – linear wetting ($M = 0.5$)	0.1	7	minutes

This puts an entirely new picture in place for the existing batch steam pressure cooking process. Uncle Tobys had originally modelled the thermal front as following the wetting front, linking them by an Arrhenius law. Now it is clear that the present process of heating and wetting concurrently is moisture-limited, and most of the hour or more of cooking time is spent waiting for moisture to penetrate the hot grain. It also means, given that the temperature is well above T_{gel}, that much of the grain is overcooked.

Already this indicates that separating the processes is a good idea, and particularly that it is a good idea to wet the grains before heating.

Mathematically, we can separate the heating and wetting processes because the timescales are so different, irrespective of whether the processes are separated in practice. That is, since the grains heat up so fast compared to the time taken to get wet, we can and do assume in the following that the temperature in the grain is constant everywhere when we model moisture penetration.

5.6 Wetting the Grains – a Nonlinear Model

A variety of models were investigated. Convection models considered the possibility of modelling moisture penetration as flow in a porous medium, driven by capillarity or swelling. These were rejected, since they do not fit with what is known about the grains. There is no air in the grain before hydration, and hence no surface tension to drive a capillary pressure. Water enters the grain by sorption processes, driven by a chemical potential called water affinity. The amount of grain swell is equal to the volume of water absorbed.

Such a process leads naturally to a nonlinear diffusion equation for moisture content. Moisture movement through whole corn kernels is considered in the paper by Syarief *et al.* ([9]) for temperatures below T_{gel}. They present the results of careful experiments on the uptake and loss of moisture from various parts of a corn kernel, and they fit the experimental results with a fully non-linear numerical finite-element model. They find that the following nonlinear diffusion equation accurately describes the uptake of moisture at 40°C:

$$\frac{\partial M}{\partial t} = \nabla \cdot (D(M)\nabla M), \tag{5.11}$$

where M is the local ratio of the mass of water to the mass of dry starch, and the nonlinearity comes from the dependence of D on M:

$$D \equiv D_0 \exp(\delta M),$$

where D_0 and δ are positive real constants. For corn starch, the values they find for D_0 and δ are tabulated above in table 5.1. The initial value M_0 of M in a "dry" grain is typically about 0.15, and the desired final moisture content, M_1, is usually about 0.6.

We nondimensionalise and re-scale equation (5.11) by choosing

$$m = \frac{M - M_1}{M_1 - M_0}, \tag{5.12}$$

$$\tau = \frac{D_0 e^{\delta M_1}}{R^2} t, \tag{5.13}$$

where R is a length scale in the problem, so that the diffusion equation becomes

$$\frac{\partial m}{\partial \tau} = \nabla \cdot (e^{\beta m} \nabla m), \tag{5.14}$$

where $\beta = \delta(M_1 - M_0)$, with initial condition $m = -1$ everywhere in the grain, and boundary condition at the grain surface $m = 0$ for $\tau > 0$. Some typical values for parameters are $\delta = 8.6$ and $\beta = 3.9$ ([9]).

This nonlinear diffusion equation (5.14) is solved numerically in section 5.6.1, and in section 5.6.2 are presented some analytical solutions in terms of the mean action time.

5.6.1 Numerical Solutions

The finite-element package called *FastFlo*™, [2], was used to solve equation (5.11) accurately with ellipsoidal geometry. Due to symmetry, only one quarter of the ellipsoid is shown. Figure 5.5 shows snapshots of the contours of constant moisture content, at times 2, 10, and 20 minutes after beginning to soak the grain. For these solutions, the value of M set at the boundary was 0.61. Contours are equally spaced in moisture content, with some values indicated directly on the plots. The contours at 10 minutes illustrate the sharpness

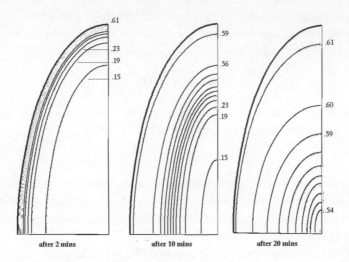

after 2 mins   after 10 mins   after 20 mins

Fig. 5.5. Numerical solutions to the nonlinear wetting problem for an ellipsoid, using *FastFlo*TM. The contours of constant moisture content are in equal increments, so only a few are labelled

of the wetting front, in the clustering near 38% moisture. After 20 minutes, most of the grain is close to 61% moisture content.

Another numerical approach used was to program a finite-difference solution using the NAG package, for a spherical grain. The sphere radius was taken to be 1.6 mm, and the diffusivity was taken as $1.5 \times 10^{-11} \exp(8.6M)$. M was set to 0.5 on the outer grain boundary. Moisture content profiles are shown in figure 5.6 at 3-minute intervals, showing the grain to be very close to equilibrium moisture levels after 21 minutes. These profiles are in good agreement with the *FastFlo*TM results.

5.6.2 Analytic Solutions – Mean Action Time

Following McNabb *et. al.* ([6], [7]), we define a mean action time at each point **r** in the grain for equation (5.14) as follows:

$$t^{**}(\mathbf{r}) \equiv \int_0^\infty \tau \left[\frac{\partial}{\partial \tau} \int_{-1}^m e^{\beta s} \, ds \right] d\tau \Bigg/ \int_0^\infty \left[\frac{\partial}{\partial \tau} \int_{-1}^m e^{\beta s} ds \right] d\tau \quad (5.15)$$

since the scaled diffusion coefficient is $e^{\beta m}$. Using integration by parts, the integrals can be simplified to give

$$t^{**}(\mathbf{r}) = \frac{\int_0^\infty (1 - e^{\beta m}) d\tau}{1 - e^{-\beta}}. \quad (5.16)$$

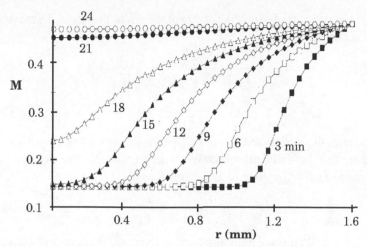

Fig. 5.6. Numerical solutions to the nonlinear diffusion equation, for a sphere, using NAG

Note that, as $\beta \to 0$, the definition in equation (5.16) is the same as equation (5.2) for the linear case.

If we define

$$\Phi \equiv \frac{1}{\beta} \int_0^\infty (1 - e^{\beta m}) d\tau, \tag{5.17}$$

so that

$$t^{**} = \frac{\beta \Phi}{1 - e^{-\beta}}, \tag{5.18}$$

then substitution into the nonlinear diffusion equation reveals that Φ satisfies a (linear) Poisson equation inside the grain:

$$\nabla^2 \Phi = -1, \tag{5.19}$$

with boundary condition on the surface of the grain $\Phi = 0$. This Poisson equation can be solved for an ellipsoidal grain with the surface given by,

$$\frac{x^2}{a^2} + \frac{y^2}{b^2} + \frac{z^2}{c^2} = 1. \tag{5.20}$$

Here, (x, y, z) and (a, b, c) are the nondimensional variables and semiaxes of the ellipsoid. The same equation for the ellipsoid applies, when (x, y, z) and (a, b, c) have dimensions. The analytic solution Φ is given in equations (5.8) and (5.9).

The largest value of Φ is at the centre of the grain. This gives the maximum mean action time, so that

$$\Phi_G = \frac{1}{2\left((1/a^2) + (1/b^2) + (1/c^2)\right)}, \tag{5.21}$$

$$t_G^{**} = \frac{\beta}{1 - e^{-\beta}}\Phi_G. \tag{5.22}$$

The term Φ_G can be seen to be a shape factor, dependent on grain shape, in the expression for the maximum mean action time t_G^{**}. The expression for maximum mean action time in *seconds* is then

$$t_G = \frac{R^2 e^{-\delta M_1}\beta\Phi_G}{D_0(1 - e^{-\beta})}, \tag{5.23}$$

provided that SI units are used throughout. Using the properties tabulated for corn starch in table 5.1, and the grain shape illustrated in figure 5.1, leads to a wetting time $t_G = 18$ minutes. Note that, while this falls between the extreme values of 7 minutes and 145 minutes calculated using linear theory in table 5.2, it is closer to the value of 7 minutes calculated using $M = 0.5$. Note also the excellent agreement with numerical results for spherical and ellipsoidal shapes.

5.6.3 Log Mean Diffusivity

A significant way to rewrite the maximum mean action time in dimensional terms is as

$$t_G = \frac{R^2\Phi_G}{D_{\text{equiv}}}, \tag{5.24}$$

where the equivalent diffusivity is the log mean of the extreme values taken by the nonlinear diffusivity,

$$D_{\text{equiv}} \equiv \frac{D_{\text{wet}} - D_{\text{dry}}}{\ln(D_{\text{wet}}/D_{\text{dry}})}, \tag{5.25}$$

where

$$D_{\text{wet}} \equiv D_0 e^{\delta M_1}, \quad D_{\text{dry}} \equiv D_0 e^{\delta M_0}. \tag{5.26}$$

Even though the moisture penetration problem is a nonlinear diffusion one, the definition used for mean action time justifies the treatment of the diffusion of moisture as an equivalent linear diffusion problem, at least in terms of the times taken to wet grains. That is, the form of equation (5.24) is the same as the time scaling obtained with a linear diffusion equation with a diffusion equal to D_{equiv}.

5.6.4 Degree of Overcook for the Present Process

The present cooking process at Uncle Tobys involves high temperature steam pressure cooking. The grains quickly establish a constant temperature, and gelatinisation depends on the penetration of the moisture front. In this section we give a formula for calculating how the degree of overcook varies with time and temperature for a single grain, for this process. The first step is to calculate the volume of grain that has been cooked.

The volume that has been gelatinised can be calculated using the mean action time to locate the wetting front as it enters an ellipsoidal grain. First we calculate the volume that is not yet cooked. After t seconds, assuming the mean action time gives the front location, the moisture front has penetrated to x, y, z values satisfying

$$t = \frac{R^2 e^{-\delta M_1} \beta \Phi(x, y, z)}{D_0(1 - e^{-\beta})} = \frac{t_G \Phi(x, y, z)}{\Phi_G}. \tag{5.27}$$

Substituting for Φ_G and Φ from equations (5.21) and (5.8) gives

$$\frac{\Phi}{\Phi_G} = 1 - \left(\frac{x^2}{a^2} + \frac{y^2}{b^2} + \frac{z^2}{c^2}\right). \tag{5.28}$$

Hence, after t seconds, all points inside the ellipsoid

$$\frac{x^2}{a^2} + \frac{y^2}{b^2} + \frac{z^2}{c^2} = 1 - \frac{t}{t_G} \tag{5.29}$$

are not yet cooked. This ellipsoid has semiaxes (Ba, Bb, Bc), where

$$B^2 = 1 - \frac{t}{t_G}. \tag{5.30}$$

The uncooked volume is the volume inside this ellipsoid:

$$\frac{4}{3}\pi abc \left(1 - \frac{t}{t_G}\right)^{3/2}. \tag{5.31}$$

Hence the volume overcooked is

$$V_{oc} = \frac{4}{3}\pi abc \left[1 - \left(1 - \frac{t}{t_G}\right)^{3/2}\right]. \tag{5.32}$$

This volume is sketched in figure 5.7. It has been normalised on the total grain volume, and the time variable has been divided by the mean action time,

$$\tilde{t} = t/t_G.$$

The degree of overcook might usefully be defined in terms of an integral overcooked grain volume,

$$\mathcal{D}_{oc} \equiv e^{T_{excess}} \int_0^{V_{oc}(t)} t \, dv, \qquad (5.33)$$

where T_{excess} is the normalised excess temperature above the gelatinisation temperature for the equilibrium moisture content of the grain, $(T - T_G)/T_G$. Transforming the integral so that everything is in terms of the time variable t gives

$$\mathcal{D}_{oc} = e^{T_{excess}} \frac{2\pi abc}{t_G} \int_0^t t \left(1 - \frac{t}{t_G}\right)^{1/2} dt, \qquad (5.34)$$

for $t \leq t_G$, and

$$\mathcal{D}_{oc} = e^{T_{excess}} \frac{2\pi abc}{t_G} \int_0^{t_G} t \left(1 - \frac{t}{t_G}\right)^{1/2} dt + \frac{4}{3}\pi abc(t - t_G). \qquad (5.35)$$

for $t \geq t_G$, when all of the grain is cooked, and it continues to overcook at a rate proportional to time. These integrals are evaluated by elementary

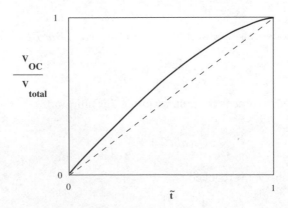

Fig. 5.7. A sketch of the calculated volume overcooked, for an ellipsoidal grain. The dashed line is a straight line of slope 1, for comparison

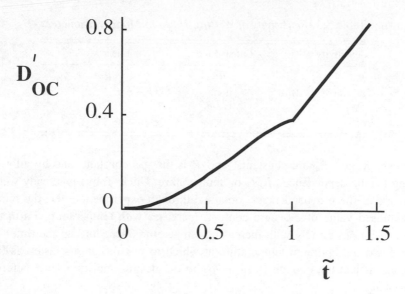

Fig. 5.8. A sketch of the normalised degree of overcook against a normalised time, for an ellipsoidal grain

methods to give in explicit form

$$
D_{oc} = e^{T_{excess}} \begin{cases} 4\pi abct_G \left[\dfrac{2}{15} + \dfrac{1}{5}\left(1 - \dfrac{t}{t_G}\right)^{5/2} - \dfrac{1}{3}\left(1 - \dfrac{t}{t_G}\right)^{3/2} \right] \\[4mm] \dfrac{8\pi abct_G}{15} + \dfrac{4}{3}\pi abc(t - t_G), \end{cases}
$$

$$(5.36)$$

the first expression applying for $t \le t_G$ and the second for $t \ge t_G$.

The way that the degree of overcook depends on time is illustrated in figure 5.8. The value plotted on the y axis is a normalised degree of overcook,

$$
D'_{oc} = \frac{\mathcal{D}_{oc}}{\frac{4}{3}\pi abct_G e^{T_{excess}}}.
$$

5.7 Temperature Dependence of Wetting Times

This is adequately incorporated into the model by taking the well-known Arrhenius dependence of D_0 on temperature T, namely

$$
D_0 = K e^{-E_a/(RT)}, \qquad (5.37)
$$

Table 5.3. *Sensitivity of Wetting Times to Model Parameters*

Sensitivity of t^{**} to:	Linear Model	Nonlinear Model
R	R^2	R^2
$\beta = \delta(M_1 - M_0)$	none	$\beta/(1 - e^{-\beta}) \sim \beta, \beta \to \infty$
D_0	$1/D_0$	$1/D_0$
M_1	none	$e^{-\delta M_1}$

where K and E_a are constants, and R is the gas constant, and by allowing for the dependence of M_1 on temperature. Uncle Tobys is already well
aware of these dependencies, obtained through experiments. M_1, the final
stabilised value of moisture content, increases with temperature. Further,
D_0 increases rapidly with increasing temperature, speeding up the time to
wet. So the optimum temperature at which to wet the grains fastest is to
select it just hot enough to raise M_1 to the desired moisture level before
cooking.

5.8 Sensitivity Analysis

The previous analytical studies lead to table 5.3, identifying the sensitivity of
the wetting time to parameters like variations in grain size, diffusivity, and
final moisture content.

5.9 Conclusions and Further Extensions

We began this study with a number of clearly specified questions, so let us
review our progress.

Various heating and wetting models have been presented and solved for
the movement of heat and moisture into whole cereal grains. By solving the
linear diffusion equation, we have shown that heat penetrates a typical grain
in a fraction of a minute, according to a simple linear diffusion model.

Moisture movement at temperatures below gelatinisation has been treated
as a nonlinear diffusion problem in the second model, leading to penetration
times of around 20–30 minutes. The concept of mean action time allows
analytic results, and leads to concise formulae for the volume of grain cooked
versus time. Also, a nice neat formula for the time to wet the whole grain is
determined.

Furthermore, estimates of degree of overcook of a grain are made, for the
present cooking process used. All models identify relevant parameters, and
give the sensitivity of cooking times to the parameters.

A major result is that the heat enters the grain on a much shorter timescale than it takes the water to enter. This result leads to the identification of a possible new way of treating the grains.

Our recommendation is that grains be presoaked at a temperature that is just below T_{gel}, so that moisture levels in the grain are as high as possible before temperature is raised above gelatinisation temperature. The problem is that the equilibrium moisture level in the grain increases with temperature, and is a little too low at lower temperatures for gelatinisation.

We estimate typical soaking times to be about 20–30 minutes. Then the wet grains can be heated at a temperature that is set at T_{gel}. The heating time should be quite short, as an individual grain is expected to heat through in a fraction of a minute. But there is more to be gained. The speed of the heating process should lead to more even cooking, because gelatinisation would occur throughout the whole grain over a very short time period. And the degree of cooking could be more precisely controlled by setting the temperature to the lowest possible to burst the starch granules. If the time required in the cookers could be reduced, plant throughput could be increased too.

Many further extensions may be made to the models discussed here. These include the following.

- A cereal grain swells significantly during wetting. Swelling can be significant in cereals. Swelling of materials as fluids are absorbed has been considered in several situations such as soil wetting and absorption of solvents in polymers. Some analysis of such a swelling model can be found in [5].

- The gelatinisation reaction and its effect on slowing moisture penetration must be considered. Discussion of the gelatinisation reaction can be found in Stapley [8]. Some analysis of this extension can be found in [5].

- The resistance effect of the pericarp on the moisture transport is apparent from NMR moisture profiles in Stapley [8]. One possible way of modelling this is found in [3].

- In the rotary cooker, an agglomeration process causes cooked product to exit in the form of globules of dough ranging from 2 to 30 cm in diameter. Considerable downstream processing is required to break down these lumps to a suitable size. The agglomeration process causes considerable compaction of product on the internal walls of the cooking vessel. This built up layer acts as a thermal insulator and hence the quality of the cooked product tends to vary with time as the thickness of the built up layer increases. Can the rate of agglomeration be predicted as a function of batch size, rotation speed, and processing time?

Acknowledgements

Thanks are due to Lyndon Ryder and Dennis Forte from Uncle Tobys for their patience and guidance during the the the 1996 Mathematics-in-Industry Study Group, University of Melbourne. Many people contributed their expertise to this problem – thanks especially to Peter McGowan, Neville Fowkes, and Colin Please.

References

[1] Carslaw, H. S. & Jaeger, J. C. (1959) *Conduction of Heat in Solids*, Oxford University Press.

[2] *FastFlo*^{TM}, (1997) Part of NAG Fortran Library. Numerical Algorithms Group Ltd, Oxford, England.

[3] Landman, K. A. & Please, C. P. (1999) Modeling moisture uptake in a cereal grain. IMA J.Math App. Buss. Ind. **10**, 265–287.

[4] McGowan, P. & McGuinness, M. (1996) Modelling the cooking process of a single cereal grain. In *Proceedings of the 1996 Mathematics-In-Industry Study Group* (ed. Hewitt, J.) University of South Australia, ISBN 0 646 28979 9, 114–140.

[5] McGuinness, M. J., Please, C. P., Fowkes, N., McGowan, P., Ryder, L. & Forte, D. (2000). Modelling the wetting and cooking of a single cereal grain. IMA J. Math. App. Buss. Ind. **11**, 49–70.

[6] McNabb, A. (1975) Asymptotic behaviour of solutions of a Stephan problem. J. Math. Anal. App. **51**, 633–642.

[7] McNabb, A. & Wake, G. C. (1991) Heat conduction and finite measures for transition times between steady states. IMA J. Appl. Math. **47**, 193–206.

[8] Stapley, A. G. F. (1995) *Diffusion and reaction in wheat grains*, PhD Thesis, University of Cambridge, England.

[9] Syarief, A. M., Gustafson, R. J. & Morey, R. V. (1987) Moisture diffusion coefficients for yellow-dent corn components. Trans. Am. Soc. Agricultural Eng. **30**, 522–528.

[10] Whistler, R. L., BeMiller, J. N. & Paschall, E. F. (1984) *Starch, Chemistry and Technology (2nd Edition)*, Academic Press, NY, USA.

Kerry A. Landman
Department of Mathematics and Statistics, University of Melbourne, Parkville, Victoria 3052, Australia
k.landman@ms.unimelb.edu.au
Mark J. McGuinness
School of Mathematical and Computing Sciences
Victoria University of Wellington, PO Box 600, Wellington, New Zealand
Mark.McGuinness@vuw.ac.nz

6

Epidemic Waves in Animal Populations:

A Case Study

Preface

The application of mathematics to biology has led to tremendous advances in the understanding of plant and animal dynamics and growth. Paramount among these areas is the spread of diseases. The application described here motivated the adaption of diffusive wave analysis to the spread of haemorrhagic disease among rabbit populations in New Zealand. The disease was introduced as an attempt, initially illegal but subsequently legalised, to control the burgeoning rabbit population in highly productive farming areas. The conceptual basis adopted was that there is a threshold maximum value of the spatial density of healthy rabbits ("susceptibles") below which the disease will not propagate. The dependence of the wave speed on the density (when it is above the threshold) can also be evaluated.

The procedure is generic and can be applied to a wide range of modelling scenarios. The nonlinear dynamics involves a range of parameters which are determined from data: the infectivity of the disease, the dispersion constant, and the infected death rate.

The purpose of this project was to see if the predicted threshold matches that seen in practice, with a view to assisting the understanding of the disease spread and the ability of it to deal with the huge problem of the endemic rabbit population many farming areas have.

The values obtained for the threshold density are close to those observed in practice and indicate that the model used is approximately correct. Of course many questions remain about the method of disease transmission. The effect of wind is taken into account. The methods employed are a healthy mix of analytical and numerical techniques, demonstrating again the interplay that underpins many successful solutions of nonlinear systems.

A major scientific study was launched in 1998 to investigate aspects of the disease, and the modelling aspects play a key role in underpinning and auditing the overall focus. This project is part of that effort.

6.1 The History of Rabbit Haemorrhagic Disease
and its Introduction into New Zealand

Rabbit Haemorrhagic Disease (RHD) is a viral disease which affects European rabbits ([29]). It first appeared in China in 1984 and has subsequently spread through most of Europe as a result of human trading of rabbits on the domestic market ([1]). Its initial appearance may have been due to the mutation of a benign strain of the virus, but this is uncertain – although there is evidence to suggest that a benign strain does exist ([3]). In 1993 the virus was introduced into Australia under quarantine by CSIRO (Commonwealth Scientific and Industrial Research Organisation) for the testing of its capabilities as a biological control of wild rabbits in both Australia and New Zealand ([17]). In 1995 a quarantined experiment began on Wardang Island but, despite precautions, the virus escaped onto the mainland and spread quickly to many other parts of Australia over the following months ([6]).

Rabbits first arrived in New Zealand with European colonists as early as 1777 ([11]). They were originally carried aboard sailing ships as a food source, but by 1866 they were regularly bred and distributed throughout parts of New Zealand's mainland and offshore islands by acclimatisation societies. They responded favourably to New Zealand's conditions. With an adequate food supply available throughout most of the year, the breeding season was lengthened; by 1889, populations had exploded, and they had become a pest.

Although it is not well quantified economically or environmentally, rabbits have a large negative impact on one of New Zealand's main industries, agriculture. Rabbit damage to New Zealand pastures has been an ongoing problem over the last century and more than $NZ600m has been spent by governments since 1950 in a bid to limit rabbit numbers ([15]). Although rabbits were previously, and in some parts of the world still are, seen as attractive pet animals, the situation of overpopulation and degradation of agricultural land had reached crisis proportions in New Zealand to the degree that desperate control measures were needed. The scientific and agricultural communities had investigated many different alternatives on how to deal with this serious problem which had potentially considerable economic threat. Biological controls were considered, and in 1987 an application to import myxomatosis virus as a biological control was rejected. Similarly, on July 2nd 1997, the application to have RHD legally introduced into New Zealand was refused by P. J. O'Hara, the Deputy Director General of MAF (Ministry of Agriculture and Forestry), ([19]). However, by August 25th 1997, there were rumours circulating that the disease was in New Zealand. This was later confirmed when the virus was found on four properties in Cromwell in the southern part of the South Island of New Zealand. After its initial release, RHD was spread mainly through the human vector. Illegally, farmers were instigating spot releases and then taking the livers of infected rabbits, grinding them in a kitchen blender and spraying this mixture over chopped carrots. The carrots were then aerially distributed over large areas. The virus was also released in the North Island and soon it appeared naturally at certain sites. It was suspected that releasing virus in this manner might be actually immunising large numbers of rabbits. It became evident that the virus was becoming established and hopes of containment or better still eradication were fading. So, in an effort to control RHD, the use of the virus as a biological control was legalised.

6.2 What is Known about the Disease

Even in the light of studies in Australia, Europe, and New Zealand, little is known about the spread of RHD. There are two possible modes of transport, the faecal/oral route ([5]) and via a possible windborne vector ([1]). The latter would explain why RHD spreads long distances in a short space of time (e.g. its escape from Wardang Island in Australia ([28])). In many Australian sites, viable RHD virus has been detected in blowflies ([24]).

Spread seems to be radial with a tendency in the prevailing wind direction in some cases ([14]). RHD epidemics occur seasonally ([29]). This is no surprise if the mechanism for the spread of the disease is via an insect vector, and also no surprise if the epidemic occurs when population densities are high, or if there is an increase of rabbit-to-rabbit contact.

6.3 What We Want to Know

From the farming perspective, the aims are:

- To minimise the adverse effects of RHD and implement management to gain full use of RHD in order to maximise the reduction of rabbit densities. For example the current percentage of kills after an epidemic of RHD is anywhere between 10% and 90% ([21]). The aim is to have a 90% kill rate in all environments.
- To find out whether RHD should be spread as a biocide (where RHD virus is spread by distribution of baits over an area) or a biocontrol (where the virus is initiated at a point source and then allowed to spread naturally).
- To determine when the virus should be released (i.e. which season), and under what conditions (temperature, rabbit density, etc.).
- If the virus is not effective, to decide on the next move.

The questions from the scientific perspective are:

- How does RHD work, and is it predictable?
- Under what conditions does it persist, and how often are epidemics? For example persistence might depend on rabbit density, time of year, temperature, or different vectors.
- Are flies the aerial vector?

Such studies are important because they may conclude that RHD as a biological control may be ineffective in certain circumstances, both in New Zealand and in other countries. Understanding a virus is crucial if it is to be used as a biological control.

6.4 The Modelling. Analytical/Numerical

We are trying to model the (natural) spread of RHD in New Zealand (and elsewhere) from a point source in order to find the critical population density threshold below which the disease will no longer persist. We can also use the model to predict the speed of the wave of infection. A specific question we addressed is how the speed of infection depends on the density of the population. A key ingredient is the need of the model outcomes to match the available data. This was the major factor in this project: that we needed to be able to predict observed outcomes and thereby have confidence in the model to predict the spread of the disease in new circumstances. This matching of model outcomes to real data is an important aspect of the modelling process.

Estimates of rabbit densities can be calculated in the field using spot light counts along a one km transect line. With this in mind we will extend an SIR (susceptibles, infectives, recovereds) model of RHD developed by [2] to incorporate a one-dimensional spatial spread of rabbits and diffusion of infection (figure 6.1). A similar approach was taken by [23], who investigated the spread of foot and mouth disease in feral pigs.

We let $I(x, t)$ be the spatial density of infectives at position x in kilometres at time t in days. We assume the point source release is at the origin $x = 0$, giving an initial condition for infectives $I(x, 0) \propto \delta(x)$, where δ is the Dirac delta function. Similarly we let $S(x, t)$ and $R(x, t)$ be the density of susceptibles and recovereds (who are immune) respectively at position x, in kilometres, and at time t, in days. We assume initially that there are no rabbits with immunity $(R(x, 0) = 0)$ and that the density of rabbits is uniform $(S(x, 0) = S_0)$.

We assume that susceptible and immune rabbits are essentially stationary, but we incorporate diffusion and advection, with parameters D and v

respectively, in the equation for the infectives. This allows for the diffusion of the infection by rabbits moving randomly and also the possibility of the infection being spread via a wind vector. Such diffusion models are derived from a random walk approach in (among others), [20], [9], [10], [25], and [16]. [8] derives the diffusion equation using the Fokker–Planck (stochastic) approach. The random walk derivation of the diffusion equation for infectives is as follows: we assume that the number of infectives at position x at time $t + \Delta t$ (given by $I(x, t + \Delta t)$) is equal to those infectives who moved from position $x + \Delta x$ a distance Δx to the left in the time interval from t to $t + \Delta t$ (i.e. $p_1 I(x + \Delta x, t)$, where p_1 is the probability of moving a distance Δx to the left) together with those infectives who were at position $x - \Delta x$ at time t and moved a distance Δx to the right during the same time interval (i.e. $p_2 I(x - \Delta x, t)$, where p_2 is the probability of moving a distance Δx to the right) added to those infectives who did not move at all from position x (i.e $(1 - (p_1 + p_2))I(x, t)$). That is,

$$I(x, t + \Delta t) = p_1 I(x + \Delta x, t) + p_2 I(x - \Delta x, t) + (1 - (p_1 + p_2))I(x, t).$$

Taylor's series expansions in Δx and Δt of the terms are taken. The equation is then rearranged and higher-order terms are ignored. When Δx and $\Delta t \to 0$, the equation becomes

$$\frac{\partial I}{\partial t} = D \frac{\partial^2 I}{\partial x^2} - v \frac{\partial I}{\partial x} \qquad (6.1)$$

where $D = \lim_{\Delta x, \Delta t \to 0}((p_1 + p_2)/2)(\Delta x^2 / \Delta t)$ and $v = \lim_{\Delta x, \Delta t \to 0}(p_2 - p_1)(\Delta x / \Delta t)$ are the diffusion and advection (constant) coefficients respectively (it should be noted that, for these limits to exist, p_1 and p_2 must depend on Δx and Δt). The actual calculation of the diffusion and advection coefficients in the field is discussed in section 6.4.3. Equation (6.1) is the standard diffusion/advection equation. Adding spread of infection, mortality due to disease, and immunity gives equation (6.3) for infectives below. Our model assumes that the rabbits can die by contracting the disease which has a constant mortality rate d per capita per day, and that we have a transmission coefficient β, which represents the proportion of susceptibles that can be infected each day by contact with one infected rabbit. Thus the spread of infection is proportional to the product of the densities of infectives and susceptibles ([13]).

We are able to ignore the spatial movement of the susceptible population as a first approximation, and so equation (6.2) is a conservation equation for susceptibles, with the term βSI being the loss due to becoming infected. We assume that both susceptibles and recovereds can breed. The breeding rate *a* of rabbits depends on pasture biomass availability ([4]) but, because the duration of infection is of the order of 40–80 days ([22]), we assume the breeding rate is constant over this time.

Similarly equation (6.4) is the conservation equation for the recovereds, where the rate of immunity is r per capita per day. Equations (6.5), (6.6), and (6.7) are initial conditions for susceptibles, infectives, and recovereds respectively. Thus, the system of equations and initial conditions governing the above assumptions is:

$$\frac{\partial S}{\partial t} = -\beta SI + a(S + R) \tag{6.2}$$

$$\frac{\partial I}{\partial t} = D\frac{\partial^2 I}{\partial x^2} - v\frac{\partial I}{\partial x} + \beta SI - dI - rI \tag{6.3}$$

$$\frac{\partial R}{\partial t} = rI \tag{6.4}$$

$$S(x, 0) = S_0 \tag{6.5}$$

$$I(x, 0) \propto \delta(x) \tag{6.6}$$

$$R(x, 0) = 0. \tag{6.7}$$

See figure 6.1 for descriptions of individual terms and table 6.1 for the units of parameters used in the model.

6.4.1 Case: No Immunity ($R(x, t) = 0$) and No Breeding ($a = 0$)

As a first step to understanding our system of equations, we assume that there is no immunity (i.e $R(x, t) = 0$) and that the release of the virus is not in the breeding season ($a = 0$). Our system becomes

$$\frac{\partial S}{\partial t} = -\beta SI \tag{6.8}$$

$$\frac{\partial I}{\partial t} = D\frac{\partial^2 I}{\partial x^2} - v\frac{\partial I}{\partial x} + \beta SI - dI \tag{6.9}$$

$$S(x, 0) = S_0 \tag{6.10}$$

$$I(x, 0) \propto \delta(x) \tag{6.11}$$

which is discussed in [16] (with $v = 0$) in terms of the spread of rabies in foxes. It is shown that the system has travelling wave solutions (figure 6.2) of the form $I(x, t) = f(x \pm ct)$ and $S(x, t) = g(x \pm ct)$.

In order to find the speed of the wave of infection analytically, [16] assumes that the rate of change of susceptibles is much slower than the rate of change

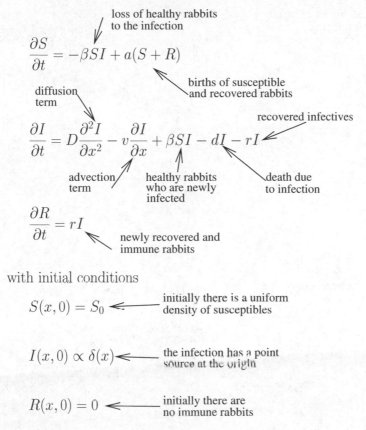

loss of healthy rabbits
to the infection

$$\frac{\partial S}{\partial t} = -\beta SI + a(S + R)$$

births of susceptible
and recovered rabbits

diffusion
term

recovered infectives

$$\frac{\partial I}{\partial t} = D\frac{\partial^2 I}{\partial x^2} - v\frac{\partial I}{\partial x} + \beta SI - dI - rI$$

advection
term

healthy rabbits
who are newly
infected

death due
to infection

$$\frac{\partial R}{\partial t} = rI$$

newly recovered and
immune rabbits

with initial conditions

$$S(x, 0) = S_0$$

initially there is a uniform
density of susceptibles

$$I(x, 0) \propto \delta(x)$$

the infection has a point
source at the origin

$$R(x, 0) = 0$$

initially there are
no immune rabbits

Fig. 6.1. Partial differential equations and initial conditions for susceptibles, infectives, and recovereds $(-\infty < x < \infty)$

Table 6.1. *Summary of Parameters*

Parameter	Units	Description
D	km^{2}day^{-1}	diffusion coefficient
v	km day^{-1}	advection coefficient
βS_0	day^{-1}	transmission coefficient scaled by initial population
r	day^{-1}	per capita recovery rate
d	day^{-1}	per capita death rate due to infection
a	day^{-1}	per capita birth rate

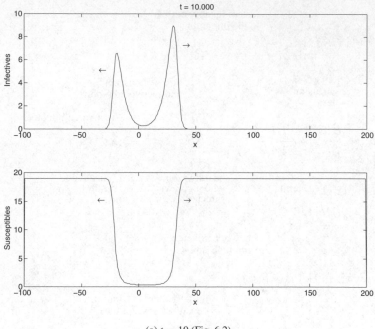

(a) $t = 10$ (Fig. 6.2)

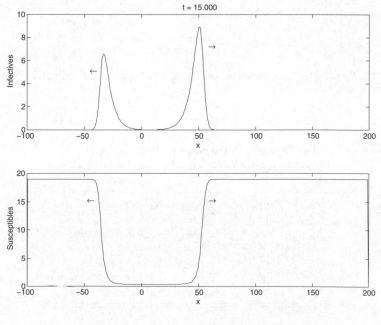

(b) $t = 15$ (Fig. 6.2)

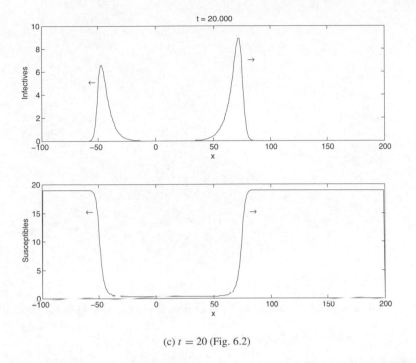

(c) $t = 20$ (Fig. 6.2)

Fig. 6.2. Travelling wave solutions of the system where there is no immunity and no breeding. The arrows indicate the direction in which the wave is travelling. Parameter values are $\beta = 0.14, d = 0.67, D = 1.5, v = 0.5, S_0 = 19, I(0,0) = 1, \Delta t = 0.1, \Delta x = 1.5\sqrt{2D\Delta t}, t_{\min} - 0, t_{\max} = 25, x_{\min} = -100, x_{\max} = 200$. The units of parameters are summarised in table 6.1

of infectives (i.e at the wave front, the density of susceptibles is approximately constant) It is then possible to solve the system of equations analytically by linearising about S_0. That is, we let $S(x, t) = S_0$ be a constant. Using Fourier transforms ([26]), we find

$$I(x, t) \sim \frac{1}{\sqrt{t}} e^{-(x-vt)^2/4Dt} e^{(\beta S_0 - d)t}$$

which is plotted in figure 6.3. Note that this is only an approximate solution of our system, and it does not have travelling waves, but we can still use it to find the speed of infection and then numerically verify this speed.

To find the speed of infection analytically, we note that, since $I(x, t) \to 0$ as $t \to \infty$, $I(x, t)$ cannot oscillate about 0. We therefore require

$$(\beta S_0 - d)t < \frac{(x - vt)^2}{4Dt} \quad \Rightarrow \quad c = \frac{x}{t} > |v \pm 2\sqrt{D(\beta S_0 - d)}|.$$

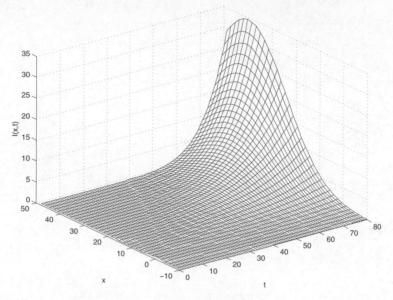

Fig. 6.3. The analytic solution for parameter values $\beta = 0.04$, $d = 0.67$, $a = 0$, $D = 1.5$, $v = 0.5$, $S_0 = 19$, $\Delta t = 1$, $\Delta x = 1.5\sqrt{2D\Delta t}$, $t_{min} = 0$, $t_{max} = 80$, $x_{min} = -100$, $x_{max} = 200$. The units of parameters are summarised in table 6.1

Therefore the centre of the wave travels with speed $c = v$, the front of the wave (travelling to the right) travels with speed $c > v + 2\sqrt{D(\beta S_0 - d)}$, and the front of the wave (travelling to the left) travels with speed $c > |v - 2\sqrt{D(\beta S_0 - d)}|$.

These analytical wave speeds ($c > v \pm |2\sqrt{D(\beta S_0 - d)}|$) have been found assuming that $\partial S/\partial t = 0$. If we drop this assumption, then we must resort to numerics to solve the system.

Explicit finite-difference numerical methods can behave pathologically when solving diffusion equations and are dependent on the step size for convergence ([7]). We therefore use an implicit method by taking the difference quotient for the second derivative and first derivatives ($\partial^2 I/\partial x^2$ and $\partial I/\partial x$) centred at (x_i, t_{j+1}) ([7], [27]), i.e

$$\frac{\partial I(x_i, t_j)}{\partial x} = \frac{I(x_{i+1}, t_{j+1}) - I(x_{i-1}, t_{j+1})}{2\Delta x},$$

$$\frac{\partial^2 I(x_i, t_j)}{\partial x^2} = \frac{I(x_{i+1}, t_{j+1}) - 2I(x_i, t_{j+1}) + I(x_{i-1}, t_{j+1})}{(\Delta x)^2}.$$

We also add boundary conditions that $I(x_{min}, t) = I(x_{max}, t) = 0$. To numerically approximate the initial condition $I(x, 0) \propto \delta(x)$, we assume that there is a density of I_0 infected rabbits in a neighbourhood of the origin.

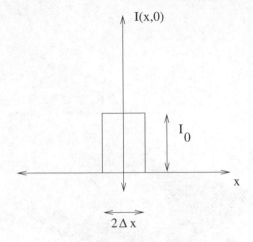

Fig. 6.4. Numerical approximation of the initial force of the infection. We assume that the density of infectives is $2\Delta x I_0$ in a neighbourhood of the origin

The force of the infection, $F(t)$, is defined as the total amount of infection present at time t; i.e. $F(t) = \int_{-\infty}^{\infty} I(x,t)dx$. Initially $(t = 0)$ the force of infection is approximated numerically (using a step size Δx) by $F(0) = \int_{-\infty}^{\infty} I(x, 0)dx = 2I_0\Delta x$ (figure 6.4).

Numerically we can see that, when $\beta \neq 0$, a wave split occurs (figure 6.5) which represents waves travelling to the left and right of the origin. In figure 6.6 we can see that wave speeds to the left and right attain their minimum, i.e. the wave speed is $c = |v \pm 2\sqrt{D(\beta S_0 - d)}|$. Note that, in figure 6.6, a negative wave speed indicated that the wave is travelling to the left. (Similarly figures 6.7 and 6.8.)

Thus we have the wave speed $c = x/t$ moving faster than the wind speed v, provided that S_0 is big enough. If $S_0 < d/\beta$, then the wave of infection will not propagate. This is the same critical density as found in [2] using a nonspatially structured model.

6.4.2 Case: No Immunity $(R(x, t) = 0)$ But Breeding Season $(a \neq 0)$

As soon as a becomes nonzero, we no longer have travelling wave solutions. Similarly the wave speed cannot be calculated as before, since we cannot make the assumption that $S(x, t)$ is constant at the wave front. However, we can see numerically (figure 6.7) that the infection at time t is travelling at a speed of approximately

$$v \pm 2\sqrt{D(\beta \times \max(S(x, t_0)) - d)}$$

where $\max(S(x, t_0))$ is the maximum number of susceptibles at time t_0.

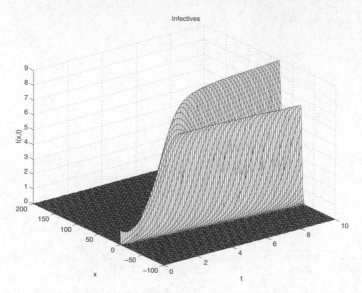

Fig. 6.5. The numerical solution of the system showing the wave split when $\beta \neq 0$. Parameter values are $\beta = 0.14$, $d = 0.67$, $a = 0$, $D = 1.5$, $v = 0.5$, $S_0 = 19$, $I(0,0) = 1$, $\Delta t = 0.1$, $\Delta x = 1.5\sqrt{2D\Delta t}$, $t_{min} = 0$, $t_{max} = 10$, $x_{min} = -100$, $x_{max} = 200$. The units of parameters are summarised in table 6.1

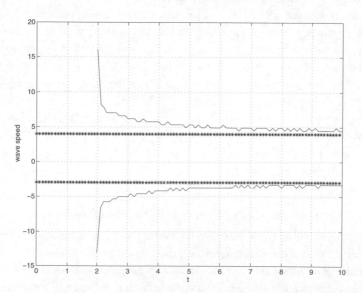

Fig. 6.6. The starred line "**" is the analytic wave speed $v \pm 2\sqrt{D(\beta S_0 - d)}$. The solid line is the numerical wave speed. The parameter values are $\beta = 0.04$, $d = 0.67$, $a = 0$, $D = 1.5$ $v = 0.5$, $S_0 = 19$, $I(0,0) = 1$, $\Delta t = 0.1$, $\Delta x = 1.5\sqrt{2D\Delta t}$, $t_{min} = 0$, $t_{max} = 10$, $x_{min} = -100$, $x_{max} = 200$. The units of parameters are summarised in table 6.1

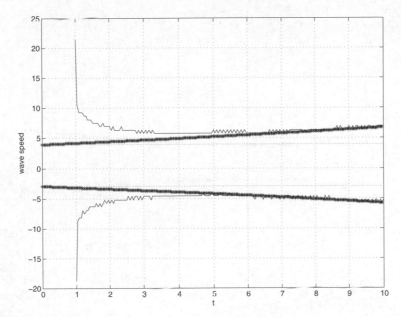

Fig. 6.7. The starred line "**" is the analytic wave speed $v \pm 2\sqrt{D(\beta \times \max(S(x, t_0)) - d)}$. The solid line is the numerical wave speed. The parameter values are $\beta = 0.14, d = 0.67, a = 0.1, D = 1.5, v = 0.5, S_0 = 19, I(0, 0) = 1, \Delta t = 0.1, \Delta x = 1.5\sqrt{2D\Delta t}, t_{min} = 0, t_{max} = 10, x_{min} = -100, x_{man} = 200$. The units of parameters are summarised in table 6.1

6.4.3 Parameter Values

Because RHD has been in New Zealand only for such a short time, there are very few data available. Scarcity of data is a common problem especially in biological modelling situations. As more data become available, we can further validate and refine our model. In the meantime we have some data from Earnscleugh station in Central Otago in the South Island of New Zealand where RHD arrived naturally. Spotlight counts prior to the arrival of RHD were 35 rabbits per 1 km of transect. After the epidemic had swept through (around about 80 days with an average speed of approximately 200 metres per day), spotlight counts had reduced to 12 rabbits per 1 km of transect. This is a reduction of about 67% [18].

The life expectancy of a rabbit which has contracted the virus is $T = 1.5$ days ([2]). Therefore the value of d, the mortality rate due to the virus, is $d = 1/T = 0.667$ per day.

The transmission coefficient β is calculated in [2] using combined data from Spain and four sites in Gum Creek Australia. The initial number of susceptibles that were infected by one infected rabbit was $\beta S_0 = 2.1$ per day.

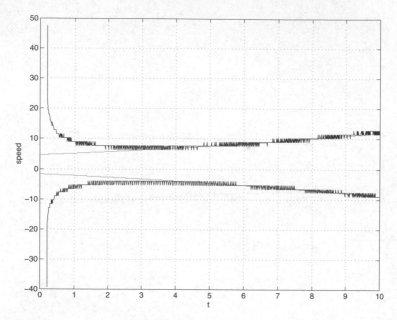

Fig. 6.8. The smooth lines are the analytic wave speeds $c(t) = v \pm 2\sqrt{D(\beta \times \max(S(x,t)) - (d+r))}$. The jagged line is the numerical wave speed. The parameter values are $\beta = 0.14$, $d = 0.67$, $a = 0.2$, $r = 0.2$, $D = 1.5$, $v = 1.5$, $S_0 = 19$, $I(0,0) = 1$, $\Delta t = 0.01$, $\Delta x = 1.5\sqrt{2D\Delta t}$, $t_{\min} = 0$, $t_{\max} = 10$, $x_{\min} = -150$, $x_{\max} = 150$. The units of parameters are summarised in table 6.1

We combine this value with the value of S_0 from the Earnscleugh site in New Zealand where densities, prior to the introduction of RHD, were 35 rabbits per km of transect. Thus $\beta = 0.06$ per day per unit density of infectives.

Using our values of d and β, we predict the critical density of susceptibles below which the disease will no longer propagate to be $d/\beta \sim 11$ rabbits per km of transect. Post-RHD densities at Earnscleugh were 12 rabbits per km of transect and there have been no further outbreaks of RHD. However d/β is extremely sensitive to values of β close to zero. Also β, which may not necessarily be constant, is sensitive to values of S_0. We clearly need further analysis of data sets in cases where RHD spread and in cases where it did not.

To find the actual speed of the wave requires the calculation of the diffusion coefficient D and the advection coefficient v. There are many ways of calculating diffusion coefficients. In [9], [20], and [10], the diffusion coefficient is calculated via the random walk derivation of the diffusion equation. [8] and [20] use the Fokker–Planck approach. Other methods are discussed (in relation to practical problems) in [20].

A way of calculating D is shown in [25] and [16] and will be discussed here. This method is practical in the field and has been used by (among others) [23], [12] and [20].

First, consider the basic 2-dimensional diffusion equation, which has a corresponding 2-dimensional diffusion coefficient D_2:

$$\frac{\partial N}{\partial t} = D_2 \left(\frac{\partial^2 N}{\partial x^2} + \frac{\partial^2 N}{\partial y^2} \right)$$

with the initial condition $N(x, y, 0) \propto \delta(x, y)$. This has the solution ([26])

$$N(x, y, t) = \frac{1}{4\pi D_2 t} \exp \frac{-(x^2 + y^2)}{4D_2 t}$$

which is the 2-dimensional normal distribution of mean $\mu = (0, 0)$ and variance $\sigma^2 = 2D_2 t$. In polar coordinates (where $r^2 = x^2 + y^2$)

$$N(r, t) = \frac{1}{4\pi D_2 t} \exp \frac{-r^2}{4D_2 t}.$$

If we think of $N(r, t)$ not as a population density function but as the probability density function that an individual is at location r at time t then $\langle r^2 \rangle$, the mean square displacement by an individual's random walk during time t, is given by

$$\langle r^2 \rangle = \int_0^\infty r^2 N(r, t) 2\pi r \, dr = 4D_2 t.$$

This gives us a way of calculating a 2-dimensional diffusion coefficient

$$D_2 = \frac{\langle r^2 \rangle}{4t}$$

To calculate a one-dimensional diffusion coefficient, we note that $D_1 = 2D_2$ ([20], [10]) and so

$$D_1 = \frac{\langle r^2 \rangle}{2t}.$$

To calculate the (one-dimensional) diffusion coefficient D for the RHD model, we let

$$D = \frac{\text{mean square displacement of a rabbit from its original location at time } t}{2t}.$$

Home ranges of rabbits vary depending on season, time of day or night, topography, distances to feeding grounds, and density of rabbits, but are approximately one hectare (100 m × 100 m) ([11]). We therefore assume that the maximum distance is 140 m per day, which gives a diffusion coefficient of $D = \frac{0.14^2}{2} = 0.0098$ km^2 per day.

The advection coefficient v in diffusion models is the drift velocity or the convective flux ([9]). In our model, v is the wind speed. If we assume that there was no wind at Earnscleugh, then we predict that the infection will travel at

$$c = 2\sqrt{D(\beta S_0 - d)} = 2\sqrt{0.0098(0.06 \times 35 - 0.667)} \sim 237 \text{ m per day.}$$

It was, however, observed at Earnsleugh that the speed of infection was on average 200 m per day ([18]) and travelled in the direction of the prevailing wind. Our speed is therefore a rough estimate. D could be too high, and we have not taken into account immunity which would slow down the speed of the wave of infection.

6.5 Immunity

There are a number of different types of immunity. First, young rabbits have natural resistance because of their age (0–8 weeks.) Second, if the mother has been exposed to the disease and survived, then the maternal antibodies are passed across the placenta and the young rabbit is immune. But this maternal immunity only lasts 10 weeks, and then the rabbit becomes susceptible. Third, adult rabbits can have antigen immunity. When they are challenged with RCD, they produce antibodies and are subsequently tested as seropositive. This immunity is for life. There could be other types of immunity: for example cellular immunity. For the purposes of modelling, we will only consider life immunity. We let r be the per capita rate of immunity in the rabbit population. The system of partial differential equations and initial conditions is that shown in figure 6.1.

Using our previous results, we can numerically verify (see figure 6.8) that the analytical wave speed is

$$c(t) = |v \pm 2\sqrt{D(\beta \times \max(S(x, t)) - (d + r))}|$$

giving a threshold density of $\max(S(x, t)) = (d + r)/\beta$.

It can be seen (figure 6.9) that, once the infection has passed through, the immune rabbits breed producing further susceptibles. The density of susceptibles that remain after an RHD epidemic may be below the threshold required for the disease to persist. Predators may keep densities low, but the need to cull immune rabbits is evident.

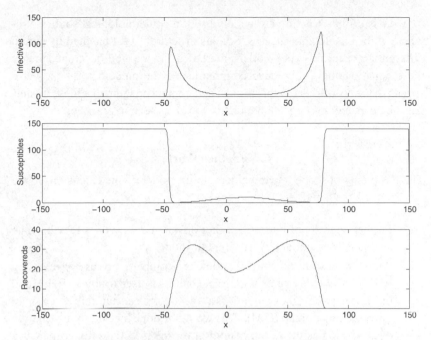

Fig. 6.9. Parameter values are the same as in figure 6.8

6.6 Results and Conclusions

This work has given a quantitative formula for the speed of the wave of infection, which depends on the density of susceptibles, the wind speed, the diffusion coefficient, the transmission coefficient, the recovery rate, and the death rate due to infection. From this it is possible to calculate the threshold density below which the infection will not spread. This threshold density is the same as that concluded in a nonspatial *SIR* model by [2]. For the calculated value of the speed to be meaningful, the parameters on which it depends must be evaluated accurately from real data. However, data adequate for this purpose are difficult to obtain in New Zealand, given the short time span that RHD has been present in the country.

The model predicts that, for the RHD virus to spread through a property, there must be a sufficiently high density of susceptibles. Thus, the best time to seed RHD would be when the entire population is challenged (i.e. when young no longer have immunity and have left the nest). After the RHD epidemic has passed through, the remaining population needs to be culled in order to eradicate any immune rabbits. The limiting factor in this study has been the paucity of data. These will take time to collect. The model developed here appears to make predictions in the range shown in the field.

Models of this structure are widely used in epidemiological studies else-where with considerable success. Spatial structure is best handled by partial differential (diffusion-type) models, thereby using the well-developed theory these equations have. We have communicated the outcomes for the wave speeds to the co-workers in this project and convinced them that these models have the capacity to replicate the diffusive waves seen in practice.

6.7 Further Work

There is plenty of scope for refinement in the model. Some extensions to the current work are listed:

- Further analysis of our work using different data sets as they become available is essential in verifying our model.

- We would also like to model and hence compare the case of releasing RHD as a biocide (where wide-scale baiting is used to spread RHD) and biocontrol (where the epidemic is allowed to spread naturally from small virus seeding points) ([21]). This would be done by using a distribution over position as the initial condition for the infectives rather than as a point source. It would be important to see how the initial profile affects the outcome for the wave speeds in the model.

- It would be important to re-examine the effect of the possible diffu-sion of the susceptible population. This would replace equation (6.3) by a diffusion equation similar to equation (6.2). The mathematical complication introduced would be considerable and would make it difficult to obtain an analytical expression for the wave speed. It is possible for some diseases that the diffusion coefficient in the suscep-tible cohort is different to that for the infective cohort. This needs to be considered.

- In Canterbury it was found that most of the survivors with immunity were females ([21]). It would be interesting to try and incorporate gender structure in our model to find out why this is the case. This would mean the model would have gender classes for both susceptible and infective cohorts adding two further diffusion–advection equations. The contact between these classes would need to be initially examined.

- We need to consider the reality of nonconstant parameter values reflecting different physical environments and climatic conditions.

All these refinements will lead to the same kind of analysis and outcomes, albeit complicated by the increased detail of the new models.

References

[1] Anti-Rabbit Research Foundation Trust (1997) Rabbit control and Rabbit Calcivirus Disease: A field handbook for land managers in Australia. Technical report, Anti-Rabbit Research Foundation Trust, 1997.

[2] Barlow, N. D. & Kean, J. M. (1998) Simple models for the impact of Rabbit Calicivirus Disease (RCD) on Australasian rabbits. Ecological Modelling **109**, 225–241.

[3] Capucci, L., Fusi, P., Lavazza, A., Pacciarini, M.-L. & Rossi, C. (1996) Detection and preliminary characterization of a new rabbit calcivirus related to Rabbit Haemorrhagic Disease Virus but non-pathogenic. J. Virology **70**, 8614–8623.

[4] Choquenot, D., Druhan, J., Lukins, B., Packwood, R. & Saunders, G. (1998) Managing the impact of rabbits on wool production systems in the central tablelands of New South Wales: an experimental study and bioeconomic analysis. In *Proceedings of the 11th Australian Vertebrate Pest Control Conference*, Bunbury, Western Australia.

[5] Coman, B. J., Staples, L. D. & Muller, J. (1994) Rabbit Haemorrhagic Disease: Issues in assessment for biological control. Chapter in: *Induction of Rabbit Calicivirus Disease (RCD) by the Oral Route in Wild Rabbits Held under Pen Conditions*. Australian Government Publishing Service, Canberra, 123–125.

[6] Cooke, B. D. (1996) Analysis of the spread of Rabbit Calicivirus from Wardang Island through mainland Australia. Technical report for the Meat Research Corporation, Australia.

[7] Cooper, J. M. (1998) *Introduction to Partial Differential Equations with Matlab*, Birkhäuser, Boston, USA, 105–109.

[8] Cox, D. R. & Millar, H.D. (1965) *The Theory of Stochastic Processes*. Methuen, Great Britain, 203–251.

[9] Ghez, R. (1988) *A Primer of Diffusion Problems*, John Wiley and Sons, New York, USA, 1–38, 85.

[10] Hoppensteadt, F.C. (1979) *Mathematical Methods of Population Biology*, New York University, New York, USA, 134–136.

[11] King, C. M. (1989) *Handbook of New Zealand Mammals*, Oxford University Press, England, 138–160.

[12] Lamoureaux, S. (1998) *Demography and population models for Hieracium pilosella in New Zealand*. PhD thesis, Agresearch (submitted).

[13] Louie, K., Roberts, M. G. & Wake, G. C. (1993) Thresholds and stability analysis of models for the spatial spread of fatal disease. IMA J. Math. Med. Bio. **10**, 207–226.

[14] Martin, G. & Donaldson, J. (1998) Rabbit control from the first RCD epidemic in New Zealand. In *Proceedings of the 11th Australian Vertebrate Pest Control Conference*, Bunbury, Western Australia, 1998.

[15] Munro, R. K. & Williams, R. T. (1994) *Rabbit Haemorrhagic Disease: Issues in assessment for biological control*. Australian Government Publishing Service, Canberra, Australia, 80–81.

[16] Murray, J. D. (1989) *Mathematical Biology*, Springer Verlag, New York, USA. Chapter 20, p697.

[17] Mutze, G., Cooke, B. & Alexander, P. (1998) The initial impact of Rabbit Haemorrhagic Disease on European rabbit populations in South Australia. J. Wildlife Diseases **34**, 221–227.

[18] Norbury, G., Heyward, R. & Parkes, J. (1998) Behaviour of Rabbit Calicivirus Disease in baited versus natural epidemics, Otago, New Zealand. In *Proceedings of the 11th Australian Vertebrate Pest Control Conference*, Bunbury, Western Australia, 1998.

[19] O'Hara, P. J. (1997) Decision on the application to approve the importation of rabbit calicivirus as a biological control agent for feral rabbits. Technical report, Ministry of Agriculture and Fisheries (MAF) New Zealand.

[20] Okubo, A. (1980) *Diffusion and Ecological Problems: Mathematical Models*, Springer-Verlag, Berlin, New York, p10, 23, 55, 75–81, 105.

[21] Otago Regional Council (1998) RCD research in Otago. Technical report, Otago Regional Council. Otago, New Zealand.

[22] Parkes, J. P., Norbury, G. & Heyward, R. (1998) Initial epidemiology of RHD in New Zealand. *New Zealand Science Review*.

[23] Pech, R. P. & McIlroy, J. C. (1990) A model of the velocity of advance of foot and mouth disease in feral pigs. J. Appl. Ecology **27**, 635–650.

[24] RCD Management Group. (1998) Australian report: National RCD monitoring and surveillance program and epidemiology program. Highlights. Technical report, RCD Management Group, Australia.

[25] Shigaseda, S. & Kawasaki, K. (1997) *Biological Invasions: Theory and Practice*, Oxford University Press, New York, 36–39.

[26] Stakgold, I. (1968) *Boundary Value Problems of Mathematical Physics*, Macmillan, New York, USA, 58–60.

[27] Twizell, E. H. (1984) *Computational Methods for Partial Differential Equations*. Ellis Horwood Limited, England, 52–253.

[28] Wardhaugh, K. & Rochester, W. (1997) Wardang Island: A retrospective analysis of weather conditions in relation to insect activity and displacement. Technical report, CSIRO Division of Entomology.

[29] Westbury, H., Lenghaus, C. & Munro, R. (1994) Rabbit Haemorrhagic Disease: Issues in assessment for biological control, chapter, *A Review of the Scientific Literature Relating to RHD*, Australian Government Publishing Service, Canberra, Australia, 123–125.

Britta Basse & Graeme C. Wake

Department of Mathematics and Statistics, University of Canterbury, Christchurch, New Zealand

b.basse@math.canterbury.ac.nz

G.Wake@math.canterbury.ac.nz

7

Dynamics of Automotive
Catalytic Converters

7.1 Introduction

Most of us know that a catalytic converter is a device that is located in our car somewhere between the engine and the tail pipe, and that its basic function is to convert harmful components of the engine exhaust, such as carbon monoxide, to less harmful ones, such as carbon dioxide. It is an important device in pollution mitigation especially in urban areas where there is a large concentration of cars. For most of us the story ends there. However, for a manufacturer of catalytic converters, such as Allied Signal, there is an obvious need to gain a deeper understanding of the behavior of catalytic converters. At the Fifth Workshop on Mathematical Problems in Industry held at RPI in May, 1989, a representative of the research group at Allied Signal presented various mathematical problems related to the dynamic behavior of catalytic converters. The industrial representative had a set of differential equations modelling the behavior of the gas concentrations and temperatures in the device but was finding it difficult to obtain numerical solutions of the equations, due, in part, to the stiffness of the chemical reaction terms which can create rapid changes in the solution. The main goal, then, was to study the set of the equations and, if possible, find reduced models and use analytical methods to predict the solution behavior and give insight into appropriate numerical techniques.

A typical automotive catalytic converter consists of an inert ceramic monolith with many narrow channels running the length of the converter, through which the exhaust gas from the engine flows (see figure 7.1). The surface of these channels is coated with a thin layer of porous proprietary material (the "washcoat"), containing various rare metals including platinum, which acts as a catalyst for the oxidation reactions that convert pollutants, such as carbon monoxide, to carbon dioxide and water. The rate at which these reactions occur, and thus the effectiveness of the catalytic converter, is highly temperature-dependent. When a car is first started and the converter is cold (at about 20°C say), the rate of reaction is small and the converter is not effective.

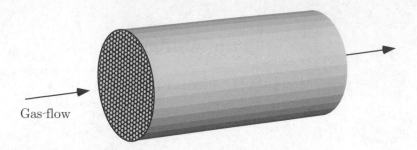

Gas-flow

Fig. 7.1. Automotive catalytic converter

As the converter heats up, there is a point somewhere along the converter where the temperature first becomes high enough for significant reaction to occur. At this point, called the light-off point, the temperature increases dramatically, and this high temperature is quickly convected downstream by the gas flow. After the light-off point forms, it moves slowly upstream towards the inlet of the converter as driven by heat diffusion in the solid. Once the light-off point reaches the inlet, the converter is in full operation.

The warm-up behavior of a catalytic converter is important because it is unable to perform its task of converting harmful carbon monoxide and unburnt hydrocarbons to carbon dioxide and water during this time, due to low converter temperatures. This warm-up period is relatively short, lasting about 5 minutes or so, but since the duration of many trips in the car is short, an increased understanding of the behavior of the catalytic converter during its warm-up period is of obvious importance.

Using numerical simulations of a rather complicated set of model equations, the industrial representative was able to determine the duration of this warm-up period. These simulations, however, were time-consuming, and it was difficult to determine the effect of the various parameters involved. In view of these difficulties, the industrial representative had two main questions:

- Can a reduced set of model equations be found whose solution describes the warm-up behavior of the catalytic converter?
- Can a solution of the reduced set of model equations be found analytically so that the dependence of the parameters on the behavior and duration of the warm-up period can be studied?

A typical trip in a car consists of several starts and stops and a variety of running speeds and engine loads. There is an initial warm-up period for the catalytic converter, as discussed, but once the converter becomes effective, it is desirable that it remains effective for the duration of the trip. Allied Signal tests the behavior of its catalytic converter designs using varying engine loads representative of a typical trip as specified by the U.S. Government. For the purpose of testing and optimizing new designs, the industrial representative had the following question:

- Can a simple and fast numerical method be found for the model equations so that the dynamic behavior of the converter can be studied for a variety of engine loads?

The first task to be addressed involves the formulation of a suitable set of equations that model the behavior of a catalytic converter. Once these equations are found, asymptotic techniques will be used to study the behavior of the solution during the warm-up phase of the converter's operation. A more detailed analysis of the equations for a variety of running conditions requires a numerical approach and this task will be addressed towards the end of this chapter. Much of the analysis discussed here originated at the RPI workshop and appears in an unpublished RPI report [2]. More detailed work was done later and is presented in a final-year undergraduate project report by Ward [7] and a paper by Please, Hagan, and Schwendeman [6].

7.2 Model Equations

The first step in developing a set of model equations involves a consideration of the system and a decision on which effects are important and should be included in the model. The main effect to consider for the catalytic converter is the chemical reactions that take place on the surface of the catalyst. These are exothermic, oxidation reactions involving carbon monoxide, unburnt hydrocarbons, hydrogen, and oxygen from the exhaust gas. The rate at which these reactions occur depends primarily on the temperature of the solid converter, as mentioned before, and on the surface concentration of the reacting species, and is controlled by the rate at which the reacting species are transferred to the catalyst surface from the gas flow. The reaction rate also depends on small concentrations of nitrogen oxides transferred from the gas and on irregularities and long-term poisoning of the catalyst surface, among others. However, these are added complications that may be considered unnecessary for a first attempt at modelling the converter behavior. Mass and heat transfer from the gas to the solid converter are important in addition to heat production in the solid due to chemical reaction. On the time scale of the model, heat diffusion in the solid is important but only in the region near the light-off point where the spatial variation in temperature is large. Heat and mass diffusion in the gas, on the other hand, are not important, because the residence time for the gas flow is too short.

There are a number of references to models of automotive catalytic converters in the literature, including Heck, Wei, and Katzer [3], Oh and Cavendish [5], Young and Finlayson [8], [9], and Zygourakis [10]. A useful starting point for the mathematical formulation of a set of model equations, and the point suggested by the representative from Allied Signal, is given by the Oh and Cavendish model. This model considers the behavior of the concentration $\bar{C}_{g,i}$ of species i in the gas and the temperature of the gas $\bar{T}_g$, and separately the corresponding concentration $\bar{C}_{s,i}$ on the surface of the catalyst, where the reactions are assumed to take place, and the temperature of the solid $\bar{T}_s$. Concentrations are given in mole fractions. Of the many chemical species present in the gas, the four oxidands CO, C_3H_6, CH_4, and H_2, indexed by $i = 1, 2, 3, 4$ respectively, and oxygen, $i = 5$, are considered to be most important. Typical values for their concentrations in the gas are 2% (CO), 450 ppm (C_3H_6), 50 ppm (CH_4), 0.667% (H_2), and 5% oxygen (see [5]). It is assumed that the concentrations and temperatures depend only on the axial distance $\bar{x}$ and on time $\bar{t}$, so that these quantities are to be interpreted as cross-sectional averages (see figure 7.2).

The equations for the gas in the Oh and Cavendish model balance convection with mass and heat transfer to the solid. These equations are

$$\epsilon \frac{\partial \bar{C}_{g,i}}{\partial \bar{t}} + \tilde{v} \frac{\partial \bar{C}_{g,i}}{\partial \bar{x}} = k_i S(\bar{C}_{s,i} - \bar{C}_{g,i}), \qquad i = 1, \dots, 5, \tag{7.1}$$

$$\rho_g c_g \left(\epsilon \frac{\partial \bar{T}_g}{\partial \bar{t}} + \tilde{v} \frac{\partial \bar{T}_g}{\partial \bar{x}} \right) = h S(\bar{T}_s - \bar{T}_g), \tag{7.2}$$

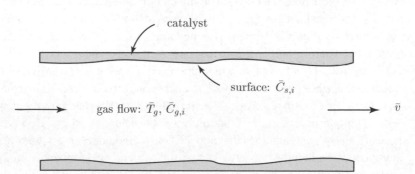

Fig. 7.2. Individual channel within the monolith

where ϵ (approximately 0.7) is the void fraction of the monolith, $\tilde{v}$ is the *apparent* gas velocity (discussed below) which may vary with $\bar{t}$ due to changing engine loads, k_i is the mass transfer coefficient of species i, S is the gas–solid surface area per unit volume of the converter, h is the heat transfer coefficient, and ρ_g and c_g are the density and heat capacity of the gas, respectively. The equations for the gas are conservation equations. For example, conservation of mass for species i in the gas implies the integral form

$$\frac{d}{d\bar{t}} \int_a^b A_o \bar{C}_{g,i} \, d\bar{x} = - \left. A_o \tilde{v} \bar{C}_{g,i} \right|_a^b + \int_a^b P k_i (\bar{C}_{s,i} - \bar{C}_{g,i}) \, d\bar{x}, \qquad (7.3)$$

where (a, b) is an arbitrary interval of $\bar{x}$, A_o is the open frontal area of the converter (the sum of the channel cross-sectional areas), $\bar{v}$ is the average velocity of the gas, and P is the perimeter of the open area. The term on the left-hand side of (7.3) represents the rate of change of mass on the interval (a, b), while the terms on the right-hand side represent the net flux of mass due to convection across the boundaries, $\bar{x} = a$ and $\bar{x} = b$, and the flux of mass to the catalyst surface. The integral form in (7.3) implies the differential conservation law

$$A_o \frac{\partial \bar{C}_{g,i}}{\partial \bar{t}} = -A_o \tilde{v} \frac{\partial \bar{C}_{g,i}}{\partial \bar{x}} + P k_i (\bar{C}_{s,i} - \bar{C}_{g,i}). \qquad (7.4)$$

In a typical experiment, the velocity is obtained from measurements of the volume flow rate q. The apparent velocity used in the Oh and Cavendish model is related to q, and the average velocity that appears in (7.3) and (7.4) by

$$q = A_o \bar{v} = A_c \tilde{v}, \qquad (7.5)$$

where A_c is the frontal area of the converter. Equation (7.1) is obtained from (7.4) by eliminating $A_o \bar{v}$ in favor of $A_c \tilde{v}$ using (7.5), dividing by A_c, and noting that $\epsilon = A_o / A_c$ and $S = P / A_c$. Similar steps involving conservation of heat energy in the gas lead to (7.2).

The Oh and Cavendish model considers the four oxidation reactions

$$CO + \tfrac{1}{2} O_2 \rightarrow CO_2$$
$$C_3 H_6 + \tfrac{9}{2} O_2 \rightarrow 3 CO_2 + 3 H_2 O$$
$$CH_4 + 2 O_2 \rightarrow CO_2 + 2 H_2 O$$
$$H_2 + \tfrac{1}{2} O_2 \rightarrow H_2 O$$

with corresponding reaction rates $\bar{R}_i$ ($i = 1, \ldots, 4$; moles per unit time per unit catalyst surface area). Reaction rates are typically complicated functions of temperature and concentration involving parameters chosen to fit experimental data. The precise forms for $\bar{R}_i$ are given in [5], but these are not essential for the present discussion. An approximate form will be discussed in section 7.3 and used in the subsequent analysis. The reaction rate is balanced by mass transfer to the catalyst surface so that

$$\bar{a}\bar{R}_i(\bar{C}_s, \bar{T}_s) = \rho_g k_i S(\bar{C}_{g,i} - \bar{C}_{s,i}), \qquad i = 1, \dots, 5, \tag{7.6}$$

where $\bar{a}$ is the normalized catalyst surface area (catalyst surface area per unit volume of the converter), which may vary with $\bar{x}$ due to design specifications or long-term poisoning. The fifth reaction rate $\bar{R}_5$ for oxygen is given by

$$\bar{R}_5 = \tfrac{1}{2}\bar{R}_1 + \tfrac{9}{2}\bar{R}_2 + 2\bar{R}_3 + \tfrac{1}{2}\bar{R}_4$$

from stoichiometry.

The final equation in the Oh and Cavendish model comes from conservation of heat energy in the solid monolith. It is assumed that the rate of change of heat energy in the solid is balanced by heat conduction in the solid, heat transfer from the gas, and heat production due to chemical reaction on the surface of the catalyst. An integral form for this balance is given by

$$\frac{d}{d\bar{t}} \int_a^b A_s \rho_s c_s T_s \, d\bar{x} = A_s K_s \frac{\partial \bar{T}_s}{\partial \bar{x}} \Big|_a^b + \int_a^b Ph(\bar{T}_g - \bar{T}_s) \, d\bar{x} + \int_a^b PQ \, d\bar{x}, \tag{7.7}$$

where $A_s = A_c - A_o$ is the solid frontal area of the converter, K_s is the thermal conductivity of the solid, Q is the rate of heat production per unit gas–solid surface area, and ρ_s and c_s are the density and heat capacity of the solid, respectively. The integral form in (7.7) implies

$$A_s \rho_s c_s \frac{\partial T_s}{\partial \bar{t}} = A_s K_s \frac{\partial^2 \bar{T}_s}{\partial \bar{x}^2} + Ph(\bar{T}_g - \bar{T}_s) + PQ, \tag{7.8}$$

assuming that ρ_s, c_s, and K_s are constants. The rate of heat production is given by the sum of the four oxidation reactions. Thus,

$$Q = \frac{\bar{a}}{S} \sum_{i=1}^4 (-\Delta H_i) \bar{R}_i(\bar{C}_s, \bar{T}_s), \tag{7.9}$$

where $-\Delta H_i > 0$ is the heat of reaction for species i (heat energy per mole). The final equation in the model is found by substituting (7.9) into (7.8) and dividing by A_c to give

$$(1 - \epsilon)\rho_s c_s \frac{\partial \bar{T}_s}{\partial \bar{t}} = (1 - \epsilon)K_s \frac{\partial^2 \bar{T}_s}{\partial \bar{x}^2} + hS(\bar{T}_g - \bar{T}_s)$$

$$+ \bar{a} \sum_{i=1}^4 (-\Delta H_i) \bar{R}_i(\bar{C}_s, \bar{T}_s). \tag{7.10}$$

The equations given in (7.1), (7.2), (7.6), and (7.10) form a complete set for the five species concentrations and temperature of the gas and the corresponding concentrations on the catalyst and the temperature of the solid. These equations are to be solved for $\bar{x}$ between 0 and L, where $L \simeq 10$ cm is the length of the converter, and for $\bar{t} > 0$ subject to various initial conditions and boundary conditions. A quick inspection of the equations reveals that the equations are linear (assuming temperature-independent coefficients) apart from the reaction rate terms, which, according to the forms given in [5], are rather complicated nonlinear functions of the surface concentrations and $\bar{T}_s$. It is these terms that present the greatest difficulty for either analytical or numerical progress.

7.3 Single-Oxidand Model and Nondimensionalization

A simplified set of model equations is introduced which retains the important elements of the full Oh and Cavendish model but is more tractable for analysis. The main simplification is to consider the behavior of a single generic oxidand, which can be regarded as carbon monoxide since it is the most abundant pollutant and the dominant heat source in the system. Within this framework, the concentrations of the remaining oxidands are neglected and thus the number of equations and complexity of the reaction law is reduced. Under normal conditions, there is a surplus of oxygen in the system which suggests a further simplification of a constant oxygen concentration. These simplifications lead to the following set of equations:

$$\epsilon \frac{\partial \bar{C}_g}{\partial \bar{t}} + \tilde{v} \frac{\partial \bar{C}_g}{\partial \bar{x}} = kS(\bar{C}_s - \bar{C}_g), \tag{7.11}$$

$$\rho_g c_g \left(\epsilon \frac{\partial \bar{T}_g}{\partial \bar{t}} + \tilde{v} \frac{\partial \bar{T}_g}{\partial \bar{x}} \right) = hS(\bar{T}_s - \bar{T}_g), \tag{7.12}$$

$$\bar{a} \bar{R}(\bar{C}_s, \bar{T}_s) = \rho_g kS(\bar{C}_g - \bar{C}_s), \tag{7.13}$$

$$(1 - \epsilon)\rho_s c_s \frac{\partial \bar{T}_s}{\partial \bar{t}} = (1 - \epsilon)K_s \frac{\partial^2 \bar{T}_s}{\partial \bar{x}^2} + hS(\bar{T}_g - \bar{T}_s)$$
$$+ \bar{a}(-\Delta H)\bar{R}(\bar{C}_s, \bar{T}_s), \tag{7.14}$$

where $\bar{C}_g$ and $\bar{C}_s$ denote the concentrations of the oxidand in the gas and on the surface of the catalyst, respectively, and the other parameters have the same meaning as in the full set of equations. For the purposes of analysis, all parameters in the equations will be regarded as constant – except for $\tilde{v}$, which will be allowed to vary with $\bar{t}$, and $\bar{a}$ which will be allowed to vary with $\bar{x}$.

Further reductions and simplifications can be made by considering a suitable nondimensional form for the equations in (7.11)–(7.14). Let

$$x = \frac{\bar{x}}{L}, \qquad t = \frac{\bar{t}}{\bar{t}_{\text{ref}}}, \qquad C_g = \frac{\bar{C}_g}{\bar{C}_{g,\text{in}}}, \qquad C_s = \frac{\bar{C}_s}{\bar{C}_{g,\text{in}}},$$

$$T_g = \frac{\bar{T}_g - \bar{T}_{g,\text{in}}}{\Delta \bar{T}}, \qquad T_s = \frac{\bar{T}_s - \bar{T}_{g,\text{in}}}{\Delta \bar{T}},$$

where $\bar{C}_{g,\text{in}} = 0.02$ and $\bar{T}_{g,\text{in}} = 300°C$ denote the oxidand concentration and gas temperature at the inlet of the converter, $\bar{x} = 0$. Obvious reference scales have been chosen for distance and concentration. There are several choices available for the reference time $\bar{t}_{\text{ref}}$, but an appropriate one for a study of the warm-up behavior is given by substituting (7.12) into (7.14) and balancing the temperature gradient $\rho_g c_g \tilde{v} (\partial \bar{T}_g / \partial \bar{x})$ due to heat transfer to the solid with the time-derivative term $(1 - \epsilon) \rho_s c_s (\partial \bar{T}_s / \partial \bar{t})$. This choice gives

$$\bar{t}_{\text{ref}} = \frac{(1 - \epsilon) \rho_s c_s L}{\rho_g c_g \tilde{v}_{\text{ref}}} \simeq 14 \, \text{sec},$$

where $\tilde{v}_{\text{ref}} = 10$ m/sec is a chosen reference value for the gas velocity. An appropriate choice for the reference temperature change is given by

$$\Delta \bar{T} = \frac{(-\Delta H) \bar{C}_{g,\text{in}}}{c_g} \simeq 300°C,$$

which is the temperature increase of the gas stream that would result from a complete reaction of all of the available oxidand. Using the dimensionless variables in (7.11)–(7.14) gives the following set of dimensionless equations:

$$\mu \frac{\partial C_g}{\partial t} + v \frac{\partial C_g}{\partial x} = \alpha (C_s - C_g), \tag{7.15}$$

$$\mu \frac{\partial T_g}{\partial t} + v \frac{\partial T_g}{\partial x} = \beta (T_s - T_g), \tag{7.16}$$

$$a R(C_s, T_s) = C_g - C_s, \tag{7.17}$$

$$\frac{\partial T_s}{\partial t} = \delta \frac{\partial^2 T_s}{\partial x^2} + \beta (T_g - T_s) + a \alpha R(C_s, T_s), \tag{7.18}$$

where the dimensionless gas velocity, normalized catalyst surface area, and reaction rate are given by

$$v = \frac{\tilde{v}}{\tilde{v}_{\text{ref}}}, \qquad a = \frac{\bar{a}}{\bar{a}_{\text{ref}}}, \qquad R = \frac{\bar{a}_{\text{ref}} \bar{R}}{\rho_g k S \bar{C}_{g,\text{in}}},$$

$\bar{a}_{ref} = 300 \text{ cm}^2/\text{cm}^3$ is a chosen reference value for the catalyst surface area per unit converter volume, and the four dimensionless parameters α, β, δ, and μ are given by

$$\alpha = \frac{kSL}{\tilde{v}_{ref}}, \qquad \beta = \frac{hSL}{\rho_g c_g \tilde{v}_{ref}}, \qquad \delta = \frac{(1-\epsilon)K_s}{\rho_g c_g \tilde{v}_{ref} L}, \qquad \mu = \frac{\epsilon L}{\tilde{v}_{ref} \tilde{t}_{ref}}.$$

Typical values for these parameters are given in table 7.1. The parameters α and β are dimensionless mass and heat transfer coefficients, respectively, and are both $O(1)$. The values for α and β listed in table 7.1 are estimates based on discussions with the industrial representative, because the values for the mass and heat transfer coefficients used in the Oh and Cavendish model are not reported in [5]. The dimensionless solid diffusivity δ is small, which indicates that heat diffusion is negligible over the time scale of interest except in regions where the gradient of the solid temperature is large. This occurs in a small region (boundary layer) about the light-off point and will be exploited in the later analysis. The last of the four parameters, μ, is the ratio of the convection time $\epsilon L/\tilde{v}_{ref}$ and the chosen reference time. This parameter is very small, indicating that the time-derivative terms in (7.15) and (7.16) may be regarded as negligible.

The reaction rates discussed in [5] are rapidly varying functions of the solid temperature. When the temperature is low ($T_s < 0$), the reaction rate is exponentially small, while at a certain temperature, called the ignition temperature, the reaction rate increases sharply. For the oxidation of carbon monoxide, this ignition temperature is close to the inlet temperature ($T_s = 0$). The reaction rate is also proportional to the surface concentration, so that the reaction turns off once the oxidand is depleted. In order to keep the analysis as simple as possible and yet still retain this basic behavior, let us consider a dimensionless reaction rate given by

$$R(C_s, T_s) = \frac{A}{\gamma} C_s e^{\gamma T_s}, \tag{7.19}$$

where γ and A are constants. In this form, γ can be regarded as a dimensionless activation energy which is typically a large value and A is an $O(1)$ dimensionless rate constant. Representative values for γ and A obtained by a rough match with the rate functions given in [5] are included in table 7.1.

Table 7.1. *Parameter Values*

$\alpha = 3$	$\beta = 3$	$\delta = .001$
$\mu = .0005$	$C_{g,in} = 1$	$T_{g,in} = -.05$
$T_{s,cold} = -1$	$A = .35$	$\gamma = 10$

For large γ, the reaction behaves like a switch about $T_s = T_{s,\text{step}} \simeq 0$ (the dimensionless ignition temperature). When $T_s < T_{s,\text{step}}$, the temperature is low and $R \simeq 0$, and when $T_s > T_{s,\text{step}}$, the temperature is high and R is exponentially large which forces $C_s \simeq 0$ according to (7.17).

Initial and boundary conditions are needed to complete the problem formulation for the single-oxidand model. At $t = 0$, the converter is cold so that

$$T_s(x, 0) = T_{s,\text{cold}} \, .$$

At the inlet, $x = 0$, the concentration of oxidand in the gas and its temperature are specified by the boundary conditions

$$C_g(0, t) = C_{g,\text{in}}, \qquad T_g(0, t) = T_{g,\text{in}} \, .$$

Finally, boundary conditions are needed for the solid temperature at both ends of the converter. Generally there would be some heat flow to the surrounding environment, but for the timescale considered the diffusion is negligible so that the no-heat-flux boundary conditions

$$\frac{\partial T_s}{\partial x}(0, t) = \frac{\partial T_s}{\partial x}(1, t) = 0$$

are a good approximation. Values for $T_{s,\text{cold}}$, $C_{g,\text{in}}$, and $T_{g,\text{in}}$ are listed in table 7.1.

7.4 Asymptotic Analysis of the Single-Oxidand Model

The dynamic behavior of a catalytic converter from a cold start to full operation has two distinct phases. In the first phase, which will be called warm-up, the converter gently heats up on an $O(1)$ timescale due mainly to heat transfer from the hot exhaust gas to the solid. Heat production due to chemical reaction is small during this phase until the solid temperature reaches the ignition temperature and light-off occurs at some point along the length of the converter. This event signals the end of the first phase and beginning of the second, which will be called light-off. Once light-off occurs, there is a rapid increase in solid temperature downstream of the initial light-off point, followed by a slow propagation of the light-off point upstream to the inlet. The second phase, which occurs on an $O(\delta^{-1/2})$ time scale, ends when the light-off point reaches the inlet and the converter is in full operation.

For the single-oxidand model, asymptotic approximations can be made to study the solution behavior during each of the two phases. A full solution requires a numerical treatment of the equations which will be discussed in the next section.

7.4.1 Warm-up Behavior

During the warm-up phase, there are no rapid spatial variations in the solid temperature, so that heat diffusion in the solid is negligible and the model equations in (7.15)–(7.18) reduce to

$$v\frac{\partial C_g}{\partial x} = \alpha(C_s - C_g), \tag{7.20}$$

$$v\frac{\partial T_g}{\partial x} = \beta(T_s - T_g), \tag{7.21}$$

$$aR(C_s, T_s) = C_g - C_s, \tag{7.22}$$

$$\frac{\partial T_s}{\partial t} = \beta(T_g - T_s) + a\alpha R(C_s, T_s), \tag{7.23}$$

where the reaction rate is given in (7.19) and the initial conditions and boundary conditions are

$$T_s(x, 0) = T_{s,\text{cold}}, \qquad C_g(0, t) = C_{g,\text{in}}, \qquad T_g(0, t) = T_{g,\text{in}}.$$

Since heat diffusion in the solid is neglected, the boundary conditions on the solid temperature are not needed. The model equations are still too difficult for analytical progress; however, one can get a qualitative picture of the solution by making further approximations.

Early in the warm-up phase, $T_s \simeq T_{s,\text{cold}}$ so that R is exponentially small. This implies that $C_s \simeq C_g$ according to (7.22) and that $C_g \simeq C_{g,\text{in}}$ according to (7.20). Further, if $T_s \simeq T_{s,\text{cold}}$ and if v is taken to be constant, then (7.21) can be integrated to give

$$T_g = T_{s,\text{cold}} + (T_{g,\text{in}} - T_{s,\text{cold}})e^{-\beta x/v},$$

which shows a decrease in gas temperature with increasing x as expected. As time passes, T_s increases from $T_{s,\text{cold}}$ due to heat transfer from the hot exhaust gas until the temperature is sufficiently large for the chemical reaction to become significant. To study this behavior qualitatively, let us consider (7.23) with a *known* heat source $B(x)$ due to heat transfer. Further let us suppose that $C_s = C_{g,\text{in}}$, so that a model problem for T_s is

$$\frac{\partial T_s}{\partial t} = B + \left(\frac{a\alpha A C_{g,\text{in}}}{\gamma}\right)e^{\gamma T_s}, \qquad T_s(x, 0) = T_{s,\text{cold}}. \tag{7.24}$$

The solution for T_s can be written implicitly in the form

$$t = \frac{T_s - T_{s,\text{cold}}}{B} - \frac{1}{\gamma B} \ln \left[\frac{\gamma B + \sigma e^{\gamma T_s}}{\gamma B + \sigma e^{\gamma T_{s,\text{cold}}}} \right], \qquad (7.25)$$

where $\sigma = a\alpha A C_{g,\text{in}}$. An example of the behavior of the solution is shown in figure 7.3. The blowup in the solution curve occurs when $t = t_b$, where

$$t_b = -\frac{T_{s,\text{cold}}}{B} - \frac{1}{\gamma B} \ln \left[\frac{\sigma}{\gamma B + \sigma e^{\gamma T_{s,\text{cold}}}} \right].$$

Such finite-time blowup (often referred to as thermal runaway) is a common feature of systems with an exponential heat source. A similar rapid increase in T_s occurs in the solution of the equations given in (7.20)–(7.23) and signals the onset of light-off. The main difference is that a constant surface concentration of oxidand is assumed in (7.24) while in (7.20)–(7.23) the solution curve for T_s would reach a plateau once the surface concentration is depleted and the reaction turns off.

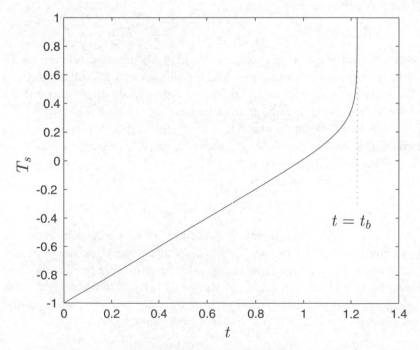

Fig. 7.3. Behavior of T_s versus time t during the warm-up phase as given by (7.25) with $B = 1$. The blowup time is $t_b = 1.225$

7.4.2 Light-off Behavior

The light-off phase is characterized by an $O(\delta^{-1/2})$ timescale progression of the light-off point towards the inlet of the converter. On either side of the light-off point, heat diffusion is negligible, and the solution has reached a steady state (on the timescale of the slow motion) so that the relevant equations for this *outer* solution are given by the warm-up phase equations in (7.20)–(7.23) with the time-derivative term in (7.23) set to zero. An *inner* solution connects the cold solid temperature upstream of the light-off point with the hot temperature downstream. The solid temperature changes rapidly across this small region, so that heat diffusion becomes important, and this causes the slow upstream motion of the light-off point.

The analysis of this phase proceeds by first considering the steady-state solution of the equations in (7.20)–(7.23). These equations are

$$v\frac{\partial C_g}{\partial x} = \alpha(C_s - C_g), \tag{7.26}$$

$$v\frac{\partial T_g}{\partial x} = \beta(T_s - T_g), \tag{7.27}$$

$$aR(C_s, T_s) = C_g - C_s, \tag{7.28}$$

$$0 = \beta(T_g - T_s) + a\alpha R(C_s, T_s), \tag{7.29}$$

where the gas velocity v is taken to be constant (or at most slowly varying). At steady state, heat transfer to the solid balances the reaction rate so that

$$\alpha(C_s - C_g) + \beta(T_s - T_g) = v\frac{\partial C_g}{\partial x} + v\frac{\partial T_g}{\partial x} = 0,$$

which implies that

$$C_g + T_g = C_{g,\text{in}} + T_{g,\text{in}}. \tag{7.30}$$

The problem reduces to solving the differential equation for C_g or T_g subject to an inlet boundary condition and the various algebraic equations. Let us concentrate on the behavior of C_g. Since

$$v\frac{\partial C_g}{\partial x} = \alpha(C_s - C_g) = -a\alpha R(C_s, T_s) < 0,$$

it is clear that C_g is a decreasing function of x. Further, (7.28) with the reaction rate in (7.19) gives

$$C_s = \frac{C_g}{1 + (aA/\gamma)e^{\gamma T_s}}, \tag{7.31}$$

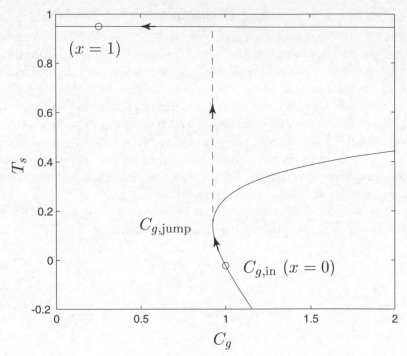

Fig. 7.4. Steady-state behavior of T_s versus C_g as given by (7.32)

which can be used in (7.29) with (7.30) to give

$$C_g = \frac{\beta \left(C_{g,\text{in}} + T_{g,\text{in}} - T_s\right) \left(1 + (aA/\gamma)e^{\gamma T_s}\right)}{\beta \left(1 + (aA/\gamma)e^{\gamma T_s}\right) - \alpha \left((aA/\gamma)e^{\gamma T_s}\right)}. \qquad (7.32)$$

The curve of T_s versus C_g as given by (7.32) has the shape of a backward S, and the portion of this curve near $C_{g,\text{in}} = 1$ is illustrated in figure 7.4. (Note that the curves at the top and bottom are connected by a lobe of the S curve which is not shown in the figure.) The solution for C_g must lie on this S curve, and the arrows indicate the path from $x = 0$ to $x = 1$. Upstream of the light-off point, C_g decreases along the curve from $C_{g,\text{in}}$ on which T_s is cold. At a point labeled $C_{g,\text{jump}}$, the curve folds back on itself and the path must jump from the cold temperature at $T_{s,\text{turn}}$ to a value on the curve where T_s is hot. The solution path continues with hot values of T_s until $x = 1$. The outer solution consists of the portion of the curve on the *cold* branch, with $C_g > C_{g,\text{jump}}$, and the portion of the curve on the *hot* branch with $C_g < C_{g,\text{jump}}$.

It is conjectured that the initial light-off point occurs at the value of x with $C_g = C_{g,\text{jump}}$ as indicated in figure 7.4. As time proceeds, the value of $C_{g,\text{jump}}$ increases slowly until the light-off point reaches the inlet and $C_{g,\text{jump}} = C_{g,\text{in}}$.

(While it has not been shown that the initial jump must occur at $C_{g,\text{jump}}$ and not sooner, the numerical evidence reported in [6] supports this conjecture.)

The speed of the light-off point is found by considering an inner problem in a small region about its position, $x = s(t)$, where

$$C_g = C_{g,\text{jump}} + o(1) \quad \text{and} \quad T_g = T_{g,\text{in}} + C_{g,\text{in}} - C_{g,\text{jump}} + o(1)$$

in the limit $\delta \to 0$. In this region, diffusion is important, and the full equation for T_s in (7.18) must be considered. Using the scaled variables

$$\xi = \frac{x - s(t)}{\delta^{1/2}}, \quad \tau = \delta^{1/2} t,$$

the equation for T_s becomes

$$-\frac{ds}{d\tau}\frac{\partial T_s}{\partial \xi} = \frac{\partial^2 T_s}{\partial \xi^2} + \beta \left(C_{g,\text{in}} + T_{g,\text{in}} - C_{g,\text{jump}} - T_s \right) + \alpha \left(C_{g,\text{jump}} - C_s \right)$$

$$(7.33)$$

to leading order, where C_s given in terms of T_s by (7.31) with $C_g = C_{g,\text{jump}}$. For a given value of $C_{g,\text{jump}}$, the speed of the light-off point, $ds/d\tau$, is found by insisting that the solution to (7.33) matches with the correct values of T_s in the outer solution as $\xi \to \pm\infty$.

It is instructive to view the inner problem in the phase plane. Let us consider a fixed position of the light-off point and a corresponding value of $C_{g,\text{jump}}$ from the outer solution. The equations in (7.33) can be written as the system

$$\begin{cases} T_s' = U \\ U' = \lambda U - S(T_s) \end{cases}, \quad (7.34)$$

where

$$U = \frac{\partial T_s}{\partial \xi}, \quad \lambda = -\frac{ds}{d\tau},$$

$$S(T_s) = \beta(C_{g,\text{in}} + T_{g,\text{in}} - C_{g,\text{jump}} - T_s) + \alpha(C_{g,\text{jump}} - C_s),$$

and the primes denote differentiation with respect to ξ. The equilibrium points occur when $U = 0$ and $S(T_s) = 0$. There are three equilibrium values of T_s corresponding to the three intersections of the curve in figure 7.4 for a given value of $C_{g,\text{jump}}$. Let us label these values $T_{s,1} < T_{s,2} < T_{s,3}$. The equilibrium points at $(T_{s,1}, 0)$ and $(T_{s,3}, 0)$ are saddle points, while $(T_{s,2}, 0)$ is an unstable node. The correct solution of the inner problem corresponds to the trajectory in the phase plane that joins the two saddle points. The problem is to find the value of λ (an eigenvalue) such that this trajectory exists. The solution of this eigenvalue problem must be determined numerically, and this will be discussed in the next section.

7.5 Numerical Methods and Results

A further study of the single-oxidand model can be made by considering numerical methods. In this section, a numerical treatment of the single-oxidand model is discussed and a solution to these equations is found. This solution illustrates the warm-up and light-off behaviors discussed previously and can be used to determine the speed of the light-off point. The speed of the light-off point as a function of its position can also be found by a numerical solution of the asymptotic equations in the limit of small δ, and this result can be compared with the speed found from the numerical solution of the full set of equations given in (7.15)–(7.18) (with μ taken to be zero).

The equations for the single-oxidand model are given in (7.15)–(7.18) with μ taken to be zero. Equations (7.15), (7.16), and (7.17) describe the spatial behavior of C_g, T_g, and C_s respectively, with the dependence on t given parametrically so that, if T_s is known at a given time, then solutions for C_g, T_g, and C_s can be found approximately by a numerical integration from the inlet at $x = 0$ to $x = 1$. For example, using (7.17) and (7.19) in (7.15) gives

$$\frac{\partial C_g}{\partial x} = -\frac{\alpha}{v}\left(\frac{(aA/\gamma)e^{\gamma T_s}}{1+(aA/\gamma)e^{\gamma T_s}}\right)C_g, \qquad \text{with } C_g(0,t) = C_{g,\text{in}}. \quad (7.35)$$

For later convenience, define

$$K(T_s, x) = \frac{(aA/\gamma)e^{\gamma T_s}}{1+(aA/\gamma)e^{\gamma T_s}}.$$

If T_s is regarded as a known function of x and t, the solution to (7.35) is

$$C_g = C_{g,\text{in}} \exp\left\{-\frac{\alpha}{v}\int_0^x K(T_s, x')\,dx'\right\}. \quad (7.36)$$

Likewise, (7.16) can be integrated to give

$$T_g = T_{g,\text{in}}e^{-\beta x/v} + \frac{\beta}{v}\int_0^x T_s e^{-\beta(x-x')/v}\,dx'. \quad (7.37)$$

The integrals in (7.36) and (7.37) can be evaluated numerically. Let $C_{g,j}$ and $T_{g,j}$ approximate C_g and T_g, respectively, at grid points $x_j = jh$, $j = 0, 1, \ldots, N$, where $h = 1/N$ for a chosen integer N (the total number of grid cells). Second-order accurate approximations for (7.36) and (7.37) are given by the iterations

$$C_{g,j} = C_{g,j-1}\exp\left\{-\frac{\alpha h}{v}K(T_{s,j}, x_{j-1/2})\right\}, \qquad C_{g,0} = C_{g,\text{in}}, \quad (7.38)$$

and

$$T_{g,j} = T_{g,j-1}e^{-\beta h/v} + T_{s,j}\left(1 - e^{-\beta h/v}\right), \qquad T_{g,0} = T_{g,\text{in}}, \qquad (7.39)$$

respectively, where $j = 1, 2, \ldots, N$ and $T_{s,j}$ approximates T_s at the cell center $(j - \frac{1}{2})h$.

The time evolution of the solid temperature on the grid is given by the ordinary differential equations

$$\frac{dT_{s,j}}{dt} = \delta\left(\frac{T_{s,j+1} - 2T_{s,j} + T_{s,j-1}}{h^2}\right) + \beta(\hat{T}_{g,j} + T_{s,j})$$

$$+ \alpha K(T_{s,j}, x_{j-1/2})\hat{C}_{g,j} \qquad (7.40)$$

for $j = 1, 2, \ldots, N$, where

$$\hat{C}_{g,j} = \frac{1}{2}\left(C_{g,j} + C_{g,j-1}\right), \qquad \hat{T}_{g,j} = \frac{1}{2}\left(T_{g,j} + T_{g,j-1}\right),$$

$$T_{s,1} = T_{s,0}, \qquad T_{s,N+1} = T_{s,N},$$

which represent a second-order accurate approximation of the evolution equation for T_s in (7.18). In view of (7.38) and (7.39), the system of equations in (7.40) has the form

$$\frac{d\mathbf{T}_s}{dt} = \mathbf{F}(\mathbf{T}_s, t), \qquad (7.41)$$

where $\mathbf{T}_s$ is a vector with components $T_{s,j}$, $j = 1, 2, \ldots, N$. The initial conditions for this system are given by $T_{s,j} = T_{s,\text{cold}}$. There are many standard numerical techniques available for integrating (7.41). The choice used here is the midpoint rule, which is a second-order-accurate Runge–Kutta method.

A numerical solution for the choice of parameters in table 7.1 is shown in figure 7.5. For this solution, it is assumed that the catalyst surface area $a(x)$ and gas velocity $v(t)$ are constant and both taken to be 1, although these functions could be allowed to vary without difficulty for the numerical method. The behaviors of C_s and T_s at successive times are shown in figures 7.5(a) and 7.5(b), respectively. The arrows in the figures indicate the progression of time. During the warm-up phase, the solid temperature is relatively cold (less than zero) and the reaction rate is negligible, so that the surface concentration of oxidand is approximately equal to the gas concentration, which, in turn, is nearly equal the inlet value $C_{g,\text{in}} = 1$. Near the end of the warm-up phase, the solid temperature approaches the ignition temperature, so that the rate of reaction becomes significant. The onset of light-off and the subsequent motion of the light-off point can be seen most clearly in the plot of C_s as the large reaction rate forces the surface concentration to zero. The solid temperature

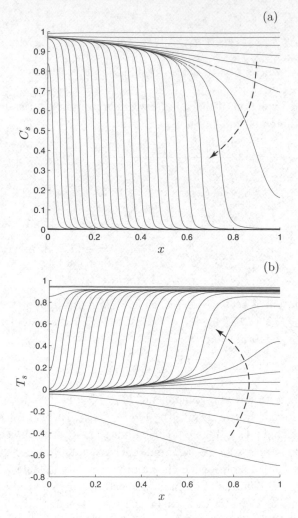

(a)

(b)

rises at the light-off point over a small $O(\delta^{1/2})$ distance, as predicted by the previous asymptotic analysis. As time proceeds, this light-off point moves slowly (on an $O(\delta^{-1/2})$ time scale) towards the inlet. The behaviors of C_g and T_g are shown in figures 7.5(c) and 7.5(d) respectively, and these curves also illustrate the predicted behaviors.

The speed of the light-off point, $ds/dt = -\delta^{1/2}\lambda$, can be found by solving the eigenvalue problem for λ as given by the system of equations in (7.34). The system of equations can be reduced to the first-order ODE

$$\frac{dU}{dT_s} = \lambda - \frac{S(T_s)}{U} \qquad (7.42)$$

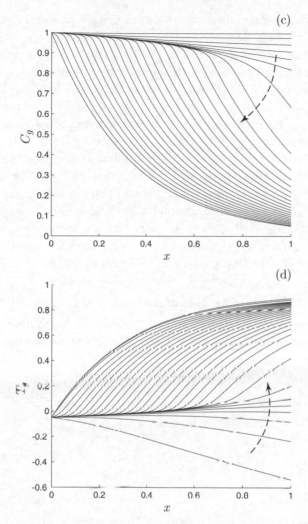

Fig. 7.5. Numerical solution for the single-oxidand model. The arrows indicate the progression of time from $t = 0$ to $t = 24$ ($\simeq$ 5.6 min). The interval of time between successive curves is 0.8 ($\simeq$ 11.2 sec)

with boundary conditions $U(T_{s,1}) = U(T_{s,3}) = 0$. For a given value of $C_{g,\text{jump}}$, this eigenvalue problem can be solved numerically using a shooting approach. The idea is to solve (7.42) for $T_s > T_{s,1}$ using the left boundary condition and separately for $T_s < T_{s,3}$ using the right boundary condition. These solutions can be found numerically using a provisional choice for λ. At the intermediate point, $T_s = \frac{1}{2}(T_{s,1} + T_{s,3})$, the values of U for the two solutions would not agree in general, and a method of iteration (Newton's method) is used to find the correct value of λ which produces agreement at this point from both sides.

Once λ is found for a given value of $C_{g,\text{jump}}$, the remaining problem is to connect the light-off position s to the corresponding value of $C_{g,\text{jump}}$. This can be done by returning to the steady-state outer equations in (7.26)–(7.29) and integrating the ODE for C_g from $x = 0$, where $C_g = C_{g,\text{in}}$, to $x = s$, where $C_g = C_{g,\text{jump}}$. For each value of s, a corresponding value for $C_{g,\text{jump}}$ is found, which, in turn, implies a value for $ds/dt = -\delta^{1/2}\lambda$ from the solution of the eigenvalue problem. The curve in figure 7.6 shows the speed of the light-off point as a function of its position for the case when $\delta = .001$. The data points in this figure show the speed of the light-off point as determined from the numerical solution of the full equations given in figure 7.5. Good agreement is noted (and expected since δ is small) except near the inlet, where the asymptotic solution is not valid, and for larger s, where there is some difficulty in identifying the light-off position as it is forming in the numerical solution.

7.6 Further Analysis of the Single-Oxidand Model

The analysis discussed in section 7.4 considered the small diffusion limit, $\delta \to 0$, which revealed the two-phase structure of the solution and the

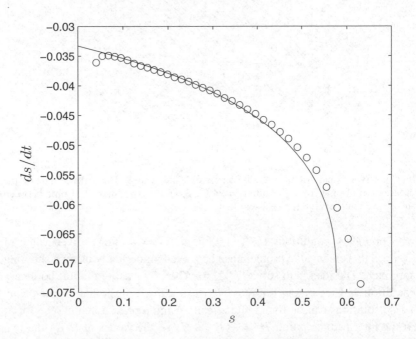

Fig. 7.6. Light-off point speed ds/dt versus position s

boundary-layer behavior near the light-off point. Analytical solutions of the reduced equations for each phase are not available (although a numerical treatment of these equations is fairly straightforward) mainly due to the non-linear reaction term which remains in these equations in its original form. In order to make further analytical progress, there needs to be some approximate treatment of the reaction term. This can be done by appealing to an asymptotic analysis of the equations in the limit of large activation energy, which implies the limit $\gamma \to \infty$ for the present reaction model. Pursuing this line of approach reveals even more structure of the equations and gives some analytical results, including an estimate for the initial position of the light-off point and a formula for the light-off point speed.

From a mathematical view, it is interesting to look at this limiting case, but in the end there will be some concern over its validity, because the large parameter in question is only 10 according to value given in table 7.1. For this reason, only a brief discussion of these results is presented here. The interested reader can find the full details in [6] or work them out independently.

An estimate for the initial light-off point comes from an analysis of the cold branch of the curve in figure 7.4. In the limit of large activation energy, it is assumed that $T_s = O(1/\gamma)$ on this branch, so that (7.32) reduces to

$$C_g = C_{g,\text{in}} + T_{g,\text{in}} - T_s + \frac{A^\star}{\gamma} e^{\gamma T_s}, \qquad (7.43)$$

where

$$A^\star = \frac{\alpha}{\beta} \left(C_{g,\text{in}} + T_{g,\text{in}} \right) a A.$$

At the inlet, $C_g = C_{g,\text{in}}$ and T_s takes a value, $T_{s,\text{in}}$ say, corresponding to the smallest solution of

$$T_{s,\text{in}} = T_{g,\text{in}} + \frac{A^\star}{\gamma} e^{\gamma T_{s,\text{in}}}.$$

It is assumed that the inlet conditions are such that $T_{s,\text{in}}$ exists on the cold branch; otherwise the only solution would lie on the hot branch of figure 7.4, and the initial light-off point would occur at the inlet. According to the reduced form in (7.43), a value for $T_{s,\text{turn}}$ is given by

$$T_{s,\text{turn}} = -\frac{1}{\gamma} \ln A^\star,$$

which corresponds to the point where $dC_g/dT_s = 0$. The values of x on this branch are determined from (7.26) which, using (7.43), reduces to solving

$$v \left(-1 + A^\star e^{\gamma T_s} \right) \frac{\partial T_s}{\partial x} = -\frac{\beta A^\star}{\gamma} e^{\gamma T_s}$$

with $T_s = T_{s,\text{in}}$ at $x = 0$. The solution is

$$\frac{1}{\gamma}e^{\gamma T_s} + A^\star T_s = -\frac{\beta A^\star}{v\gamma}x + \frac{1}{\gamma}e^{\gamma T_{s,\text{in}}} + A^\star T_{s,\text{in}}.$$

The initial light-off point, $x = s_0$ say, occurs when $T_s = T_{s,\text{turn}}$, which gives the estimate

$$s_0 = \frac{v}{\beta}\left(\ln A^\star - 1 + \gamma T_{s,\text{in}} + \frac{1}{A^\star}e^{\gamma T_{s,\text{in}}}\right).$$

It is interesting to note that, if the inlet conditions are such that $s_0 > 1$, then the system would be too cold for light-off to occur.

An estimate for the speed of the light-off point can be found by considering the eigenvalue problem given by

$$-\frac{ds}{d\tau}\frac{\partial T_s}{\partial \xi} = \frac{\partial^2 T_s}{\partial \xi^2} + \beta\left(C_{g,\text{in}} + T_{g,\text{in}} - C_{g,\text{jump}} - T_s\right) + \alpha\left(C_{g,\text{jump}} - C_s\right)$$

(7.44)

in the limit $\gamma \to \infty$. The value of C_s is given by (7.31) with $C_g = C_{g,\text{jump}}$. The simplest analysis of this problem replaces (7.31) by the step function

$$C_s = C_{g,\text{jump}} H\left(T_{s,\text{step}} - T_s\right),$$ (7.45)

where H is the Heaviside function and $T_{s,\text{step}}$ is the ignition temperature. This approximation has the correct behavior away from the light-off point and would give the correct speed for the some value of $T_{s,\text{step}}$. While it is not known what the correct value is, a reasonable choice is

$$T_{s,\text{step}} = -\frac{1}{\gamma}\ln\left(\frac{aA}{\gamma}\right),$$

which corresponds to the value where the reaction rate changes most rapidly. This choice is used in [6] and suggested by the work of [1] and [4]. It is straightforward to solve the linear equation in (7.44) with C_s given by (7.45) for T_s on either side of $T_{s,\text{step}}$ and then join the solutions at $T_{s,\text{step}}$. This results in an estimate for the light-off point speed given by

$$-\frac{ds}{d\tau} = \frac{\alpha C_{g,\text{jump}} - 2\beta\Gamma}{\left(\alpha C_{g,\text{jump}} - \beta\Gamma\right)^{1/2}\Gamma^{1/2}},$$

where

$$\Gamma = T_{s,\text{step}} - T_{g,\text{in}} + C_{g,\text{jump}} - C_{g,\text{in}},$$

which reduces to

$$-\frac{ds}{d\tau} = \left(\frac{\alpha C_{g,\text{jump}}}{\Gamma}\right)^{1/2} \tag{7.46}$$

upon noting that Γ is small and positive and $C_{g,\text{jump}} = O(1)$.

A formal asymptotic analysis can be made of the equations in (7.44) and (7.31) without the use of the step-function approximation. The structure of the solution is more complicated than the previous analysis. Four regions emerge with appropriate scalings of ξ and T_s for each. A solution can be found for the relevant equations in each region and then matched with the solutions on either side. Details are given in [6]. In the end, the analysis yields a formula for the light-off point speed which agrees to leading order with the one given by (7.46).

The estimates for the initial position of the light-off point and its speed can be checked by comparing these results with the numerical solution of the equations. This is done in [6], and it is found that the estimate for the initial light-off position is in good agreement with the numerical results, while the speed of the light-off point is not. The asymptotic expansion for the $ds/d\tau$ involves powers of the "small parameter" $1/\ln\gamma = 0.43$, which is not very small, and would suggest poor accuracy for an estimate using only the leading terms in the expansion.

7.7 Concluding Remarks

A detailed asymptotic and numerical treatment has been given for the single-oxidand model with the aim of answering the questions posed by the industrial representative. Let us review these questions briefly and summarize the results that have been obtained. The first question concerned the formulation of a reduced model suitable for analysis. This question has been answered with the formulation of the single-oxidand model. This model is significantly simpler than the original Oh and Cavendish model and yet retains the essential features of the more complicated model. The second question involved an analysis of the reduced model and a search for analytical solutions to study the warm-up period of the converter's operation. An analysis of the reduced model in the limit of small diffusion revealed a two-phase structure during the warm-up period. The timescale for the first (warm-up) phase was found to be $O(\bar{t}_{\text{ref}})$ while the timescale for the second (light-off) phase was found to be much slower. The timescale for the latter phase was found to be $O(\delta^{-1/2}\bar{t}_{\text{ref}})$. A good understanding of the solution was obtained analytically for each phase, but a numerical method was needed to find the solutions quantitatively. A relatively simple numerical method was found for the reduced model. This numerical method was used to verify analytical results, but could also be used to find solutions for varying running conditions or for a variety of parameter values in answer to the last question posed by the industrial representative.

The work presented here can be extended in various ways. For example, the simple reaction given in (7.19) for the reduced model could be replaced by one that includes the effects of oxygen depletion and nitrogen oxide inhibition. This extension brings the reaction rate model closer to the more complicated forms given in [5] and would require additional equations for the gas and surface concentrations of oxygen. The concentration of nitrogen oxides is small and could be taken to be constant. This extension is discussed in [6]. A further extension, discussed in [7] following the work in [10], considers a two-dimensional model that includes radial variations in the converter cross-section due to nonuniformities in gas velocity and solid temperature. A fairly straightforward modification of the numerical method discussed here could be used to study these radial variations.

Acknowledgements

The initial work on this problem was carried out at the Fifth RPI Workshop on Mathematical Problems in Industry, and I would like to thank John Nelligan for bringing the problem and the many participants at the workshop for their collaborative effort. Further polishing of the work was done after the workshop, and I thank Colin Please and Pat Hagan for their efforts.

References

[1] Byrne, H. & Norbury, J. (1990) Catalytic converters and porous medium combustion. In *Emerging Applications in Free Boundary Problems, Proc. 5th International Colloquium on Free Boundary Problems* (eds. Chadam, J.M. & Rasmussen, H.) Montreal, Canada, May 1990.

[2] Hagan, P. S., King, J. R., Nelligan, J. D., Please, C. P. & Schwendeman, D. W. (1989) Dynamic behavior of automotive catalytic converters. In *Proc. 5th RPI Workshop on Mathematical Problems in Industry*, Troy, New York, USA.

[3] Heck, R. H., Wei, J. & Katzer, J. R. (1976) Mathematical modeling of monolithic catalysts. AIChE J. **22**, 477–484.

[4] Norbury, J. & Stuart, A. M. (1989) Models for porous medium combustion. Quart. J. Mech. Appl. Math. **42**, 159–178.

[5] Oh, S. H. & Cavendish, J. C. (1982) Transients of monolithic catalytic converters: Response to step changes in feed-stream temperature as related to controlling automotive emissions. Ind. Chem. Prod. Res. Dev. **21**, 29–37.

[6] Please, C. P., Hagan, P. S. & Schwendeman D. W. (1994) Light-off behavior of catalytic converters. SIAM J. Appl. Math. **54**, 72–92.

[7] Ward, J. P. (1991) *Mathematical modeling of the dynamics of monolithic catalytic converters for the control of automotive emissions*, Mathematics and Computer Science Final-Year Project, University of Southampton, England.

[8] Young, L. C. & Finlayson, B. A. (1976) Mathematical models of the monolith catalytic converter: Part I. Development of model and application of orthogonal collocation. AIChE J. **22**, 331–343.

[9] Young, L. C. & Finlayson, B. A. (1976) Mathematical models of the monolith catalytic converter: Part II. Application to automobile exhaust. AIChE J. **22**, 343–353.

[10] Zygourakis, K. (1989) Transient operation of monolith catalytic converters: A two-dimensional reactor model and the effects of radially nonuniform flow distributions. Chem. Eng. Sci. **44**, 2075–2086.

Donald W. Schwendeman

Department of Mathematical Sciences, Rensselaer Polytechnic Institute, Troy, New York, 12180-3590, USA

schwed@rpi.edu

8

Analysis of an Endothermic Reaction
in a Packed Column

8.1 Introduction

In 1977 the British Steel Corporation presented a problem to the European Study Groups for Industry, concerning the analysis of a gas–solid reaction in a column packed with solid catalyst pellets. This is an industrial problem of common concern, arising in various kinds of reactor such as a blast furnace. BSC were coy about the details of the reaction involved, but were happy to provide details of relevant physical parameters, essential for any analysis, and this allowed for a fairly thorough study of the problem.

Much of the applied mathematical interest in this type of problem has been motivated by the occurrence of *exothermic* reactions, i.e. those which release heat, as multiple steady states may exist for such reactions, and the associated phenomenon of thermal runaway can occur. However, the BSC problem concerned an *endothermic* reaction, one which absorbed heat, and our challenge thus lay in seeking some analytical understanding of the dynamics of the reaction.

The specific problem as summarised by BSC was as follows:

The problem is to solve two simultaneous partial differential equations. Although a method has been found using finite differences, the method is very slow in convergence and the boundary conditions have to be solved by an indirect manner. This is because the boundary conditions are dependent on an external radiative heat transfer problem. It would be very useful if an alternative approach to the problem was available in which the external heat transfer problem and the partial differential equations were formulated as a single problem.

Then, as now, the industry really wanted an efficient numerical code; and then, as now, our approach was directed towards analytical understanding. For a problem as simple as that stated below, it is in fact feasible to provide a good description of the solution analytically; furthermore, a numerical solution requires a certain amount of analytical pre-treatment, to know what time and space steps should be used. In a sense, the challenge that this problem avoids raising, but which is also a suitable case for treatment, is the industrially relevant situation where ten, twenty, or a hundred reactions are going on. If we judge the state of the art by the extant literature (which is very substantial), then we find that much of the modelling and analysis that has been done is limited to the consideration of single-pellet dynamics, and also simple first- or second-order reaction kinetics (for examples of the gas–solid reaction literature, see [1], [3]–[6], [9]–[16], [18]–[22], [24], and [25]). It is thus perhaps a message yet to be realised, that applied analytic techniques can not only facilitate the solution of relatively simple problems such as that here, but are also capable of dealing with more realistic problems of practical concern.

8.2 The Problem and the Model

A cylindrical tube, of typical radius 10 cm and height 8 m, is packed with solid catalyst pellets, which react with a gas stream flowing through the tube. The situation is shown in figure 8.1.

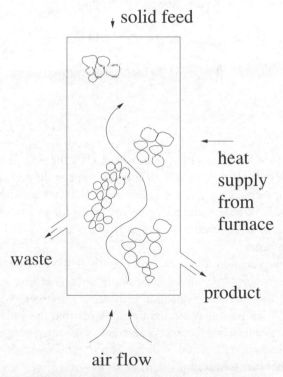

Fig. 8.1. A schematic illustration of the reactor geometry

A first-order reaction occurs between the gas and the solid, and the reaction is endothermic. If the solid pellets are sufficiently small, then the solid pellet temperature s may be taken to be locally uniform (within each pellet). It is always the case that the pellets are much smaller than the reactor radius, and it is commonly also the case that the pellets are in thermal equilibrium, and at relatively uniform temperature. In the present instance, this is confirmed *post hoc* by the fact that the macroscopic heat conduction parameter Γ is $O(1)$.

The gas (at mean temperature g) supplies heat to the solid through a viscous boundary layer (in the gas, next to the solid surface). For a typical gas velocity of 100 cm s^{-1}, pellet radius 1 cm, and gas viscosity 1 cm^2 s^{-1} (at reactor temperatures), the particle Reynolds number is 100, so that the flow would be laminar, although the irregular pore space is liable to make it highly unsteady. The Prandtl number of air at elevated temperatures is approximately one, so that the Peclet number is $\sim 10^2$, and a thermal boundary layer (slightly thicker than the viscous one) will exist next to the solid surface. It is not in fact essential that these boundary layers exist, since the model below will apply in any case, even if the particle Peclet number is small.

A steady-state model to describe this situation is

$$\rho_g c_p u \frac{\partial g}{\partial x} - h_v(s - g) = k_g \nabla^2 g,$$

$$h_v(s - g) + \mathcal{R} \Delta H = k_s \nabla^2 s,$$

$$u \frac{\partial c}{\partial x} = -\mathcal{R},$$

$$\mathcal{R} = K c e^{-\lambda/s}, \tag{8.1}$$

wherein u is the gas flux, ρ_g the gas density, and c_p the gas specific heat; h_v is a heat transfer coefficient (whose size will depend on the particle Reynolds and Peclet numbers), k_g and k_s are the effective thermal conductivities of gas and solid, and $\mathcal{R}$ is the reaction rate, given by $(8.1)_4$ for a first-order reaction; c is the gas reactant concentration, and ΔH is the heat of reaction absorbed per mole of reactant.

The first two equations represent heat transfer in the gas and solid phases. The terms in the gas temperature equation describe heat advection, interfacial heat transfer, and heat conduction respectively. In the second equation, the heat advection term is neglected on the basis that the solid velocity is negligible, so that heat conduction is balanced by the interfacial heat transfer and also the endothermic heat of reaction $\mathcal{R} \Delta H$. This term affects the solid temperature because the reaction occurs at the gas–solid interface. Depending on the nature of the pellet, this is either the pellet surface or, if the pellets are porous, the internal surface area. In either event, the heat of reaction is removed from the solid.

The third equation represents the depletion of reactant in the gas stream. The rate of reaction $\mathcal{R}$ acts as a volume sink for the reactant, while the effect of diffusion can be ignored, being small. The final equation is the Arrhenius term for the rate of a first-order reaction, dependent on the solid (surface) temperature.

A good deal of physics is present in these equations, and we shall have some comments to make concerning the validity of the model in the discussion. For now we focus our attention on the model (8.1), supplemented by boundary conditions as indicated below. The inlet and outlet conditions are controlled by the rapid gas flow through the tube. In general, continuity of heat flux across a boundary has two components, due to advective flux and conductive heat transfer. When the advective term is large (as we suppose here), this essentially means that the temperature will be continuous. At the inlet, the gas flow carries in its external temperature, which we therefore consider prescribed, thus

$$g = s = g_0, \quad c = c_0 \quad \text{on} \quad x = 0, \tag{8.2}$$

where in addition we prescribe the entry reactant concentration of the gas stream. At the outlet, the same principle applies, except that the gas carries its temperature out of the tube. In the absence of any external control at the outlet, it is natural to suppose that there is no thermal boundary layer at the outlet, and a mathematically consistent way of ensuring this is to prescribe

$$\frac{\partial s}{\partial x} = \frac{\partial g}{\partial x} = 0 \quad \text{on} \quad x = L, \tag{8.3}$$

although further consideration of the validity of this might be necessary in differing circumstances. In the present case, the outlet conditions will not actually affect the discussion.

Finally, on the wall,

$$k_g \frac{\partial g}{\partial r} = h_0(T - g),$$

$$k_s \frac{\partial s}{\partial r} = \varepsilon_G \sigma (T^4 - s^4). \tag{8.4}$$

These Neumann boundary conditions differ in form because it is assumed that heat transfer to the gas from the wall at temperature T is by thermal conduction, whereas transfer to the solid pellets occurs through radiation; σ is the Stefan–Boltzmann constant and ε_G is the emissivity.

We nondimensionalise these equations by writing

$$g = g_0 + (T - g_0)g^*, \quad s = g_0 + (T - g_0)s^*,$$

$$x = r_0 x^*, \quad r = r_0 r^*, \quad c = c_0 c^*, \tag{8.5}$$

where r_0 is the tube radius. It is common in combustion theory to linearise the exponent in the Arrhenius exponential term (this is called the Frank–Kamenetskii approximation). The asymptotic basis on which this assumption ultimately rests is that the temperature variation in the problem is much less than the "activation" temperature λ (actually this is equal to the "activation energy" divided by the universal gas constant). In the example we use below, $\lambda = 5000$ K, whereas the external furnace temperature is 1300 K, and the inlet gas temperature is 700 K; thus the maximum temperature range is 600 K, much less than λ. The formal derivation requires

$$(T - g_0)s^* \ll g_0, \tag{8.6}$$

so that when the exponent is expanded to two terms of its Taylor series, we obtain

$$e^{-\lambda/s} \approx e^{-\lambda/g_0} e^{\Lambda s^*}, \tag{8.7}$$

where

$$\Lambda = \frac{\lambda(T - g_0)}{g_0^2}. \tag{8.8}$$

Apparently the assumption (8.6) is not supported by the values $T = 1300$ K and $g_0 = 700$ K, but in fact we later find that $s^* \sim 1/\Lambda$, so that (8.6) is equivalent to $g_0 \ll \lambda$, which *is* valid. The model (8.1) can then be written in the dimensionless form, dropping the asterisks,

$$\alpha \frac{\partial g}{\partial x} - (s - g) = \mu \nabla^2 g,$$

$$h(s - g) + ce^{\Lambda s} = \Gamma \nabla^2 s,$$

$$\frac{\partial c}{\partial x} = -\kappa c e^{\Lambda s}, \tag{8.9}$$

and the boundary conditions are

$$c = 1, \quad g = s = 0 \quad \text{on} \quad x = 0;$$

$$\frac{\partial g}{\partial x} = \frac{\partial s}{\partial x} = 0 \quad \text{on} \quad x = 1/\delta;$$

$$\gamma \frac{\partial g}{\partial r} = 1 - g,$$

$$\beta \frac{\partial s}{\partial r} = 1 - (\beta_1 + \beta_2 s)^4 \quad \text{on} \quad r = 1. \tag{8.10}$$

The parameters α, μ, h, Γ, κ, δ, γ, β, β_1, and β_2 are given by the values

$$\alpha = \frac{\rho_g c_p u}{h_v r_0}, \quad \mu = \frac{k_g}{h_v r_0^2},$$

$$h = \frac{h_v(T - g_0)}{\theta \Delta H}, \quad \Gamma = \frac{k_s(T - g_0)}{r_0^2 \theta \Delta H},$$

$$\kappa = \frac{\theta r_0}{u c_0}, \quad \delta = r_0/L,$$

$$\gamma = \frac{k_g}{h_0 r_0}, \quad \beta = \frac{k_s(T - g_0)}{r_0 \varepsilon_G \sigma T^4}, \quad \beta_1 = \frac{g_0}{T}, \quad \beta_2 = 1 - \beta_1, \tag{8.11}$$

where

$$\theta = K c_0 e^{-\lambda/g_0}. \tag{8.12}$$

Typical values of the parameters supplied by BSC were as follows:

$$\Delta H \sim 5.7 \times 10^4 \text{ cal mole}^{-1},$$

$$K \sim 0.9 \times 10^4 \text{ s}^{-1}, \quad \rho_g u \sim 1.9 \times 10^{-3} \text{ mole cm}^{-2} \text{ s}^{-1},$$

$$\lambda \sim 5 \times 10^3 \text{ K}, \quad c_0 \sim 0.4 \times 10^{-5} \text{ mole cm}^{-3},$$

$$u \sim 100 \text{ cm s}^{-1},$$

$$c_p \sim 0.2 \text{ cal mole}^{-1} \text{ K}^{-1}, \quad T \sim 1300 \text{ K}, \quad g_0 \sim 700 \text{ K},$$

$$r_0 \sim 10 \text{ cm}, \quad h_v \sim 2.3 \times 10^{-3} \text{ cal cm}^{-3} \text{ s}^{-1} \text{ K}^{-1},$$

$$k_g \sim 5 \times 10^{-4} \text{ cal cm}^{-1} \text{ s}^{-1} \text{ K}^{-1}, \quad k_s \sim 0.1 \text{ cal cm}^{-1} \text{ s}^{-1} \text{ K}^{-1},$$

$$h_0 \sim 5 \times 10^{-4} \text{ cal cm}^{-2} \text{ s}^{-1} \text{ K}^{-1}, \quad \varepsilon_G \sim 0.8,$$

$$\sigma \sim 1.37 \times 10^{-12} \text{ cal cm}^{-2} \text{ s}^{-1} \text{ K}^{-4}, \quad L \sim 800 \text{ cm}. \tag{8.13}$$

From these we find typical values of the constants to be

$$\alpha \sim 1.65 \times 10^{-2}, \quad \mu \sim 2.2 \times 10^{-3}, \quad \theta \sim 2.85 \times 10^{-5} \text{ mole cm}^{-3} \text{ s}^{-1},$$

$$h \sim 0.85, \quad \Gamma \sim 0.37, \quad \kappa \sim 0.72, \quad \delta \sim 1.25 \times 10^{-2},$$

$$\Lambda \sim 6, \quad \gamma \sim 0.1, \quad \beta \sim 1.9, \quad \beta_1 \sim 0.54, \quad \beta_2 \sim 0.46. \tag{8.14}$$

The presence of several small parameters suggests a variety of possible asymptotic limits, but before we proceed to their consideration, we append some comments on the model as supplied by the company.

Industrial Specification

The model as presented by BSC was not quite in the above form. The sign of the h_v term in (8.1)$_1$ was positive, the mass flow was given as $M = \rho_g u$, and the reaction equation for c was absent, with the heat absorption $\mathcal{R}\Delta H$ being given as $Q = Ae^{-\lambda/s}$. Presumably (it *was* over twenty years ago) the reaction equation for c was added during the study group, and a relevant value for ΔH was then applied. Presumably the gas is air and the gaseous reactant is oxygen (21% by volume). At 400°C, the density of air is about 0.5×10^{-3} gm cm^{-3}, and with its molecular weight being about 30 (gm mole^{-1}), this is a molar density of about 1.65×10^{-5} mole cm^{-3}, and 0.21 of this is 0.35×10^{-5} mole cm^{-3}, consistent with the supplied value. On the other hand, $M = \rho_g u$ must be the mass flow rate of air, so that the molar density ρ_g is about $5c_0$. If we use the supplied value of c_0, then we derive $u \sim 100$ cm s^{-1}. In fact, in the original report, a value of $u \sim 450$ cm s^{-1} was used, which reduces the value of κ by a factor 4.5. Presumably this oversight was due to neglect of what the gaseous reactant actually was.

Another comment worth making is that the heat transfer coefficient h_v and thermal conductivity k_s and k_g should be phase-averaged in some way (the superficial velocity u (i.e. the volume flux per unit area) is already effectively phase-averaged); no information is available as to whether this was in fact done, although it would only make cosmetic differences to the results.

The other parameters were supplied by BSC, though with some inaccuracies: r given as 1.38×10^{-2}, mass flow M as mole cm^{-2}.

8.3 Analysis

The original presentation of this problem was concerned, as is often the case, with the numerical solution of the model. The company had not nondimensionalised the model, and one can infer that the nature of their concern was such that the accurate solution of this *particular* model was perhaps less relevant than a qualitative understanding of how the model ought to be solved. In particular, the choice of a first-order reaction is presumably a gross simplification of the realistic chemistry. In this context, it ought to be pointed out that the inclusion of detailed chemistry is likely to lead to other, faster, reactions with associated short relaxation distances, and this will also cause difficulties in straightforward numerical schemes.

We now proceed to an analysis of the model. The parameter Λ is relatively large, particularly as it is present in an exponent. If we suppose that s is negative (and $O(1)$), then the heat absorption term is exponentially small and negligible. However, there is then no mechanism which can reduce s to such low values, and this assumption seems unwarranted. In a similar vein, if s is positive and $O(1)$, then the exponential term is dominant, suggesting to leading order either that $c \approx 0$ (which cannot be valid near the inlet), or that s is small and negative. And if this is the case, it seems likely that g also will be small and negative. This provides the motivation to rescale the temperatures to satisfy

$$g \sim s \sim \frac{1}{\Lambda}, \tag{8.15}$$

so that the corresponding dimensional temperature scales are g_0^2/λ. Of course we might have scaled this way initially, but it is more illuminating of the solution process to indicate the rescalings as they are discovered.

Adopting (8.15), the (rescaled) model is

$$\alpha g_x - (s - g) = \mu \nabla^2 g,$$

$$H(s - g) + \frac{c}{\varepsilon} e^s = \nabla^2 s,$$

$$c_x = -\kappa c e^s, \tag{8.16}$$

where

$$H = h/\Gamma \sim 2.2, \quad \varepsilon = \Gamma/\Lambda \sim 0.06, \tag{8.17}$$

with the side wall conditions being

$$\frac{\gamma}{\Lambda} \frac{\partial g}{\partial r} = 1 - \frac{g}{\Lambda},$$

$$\frac{\beta}{\Lambda} \frac{\partial s}{\partial r} = 1 - \left(\beta_1 + \frac{\beta_2 s}{\Lambda} \right)^4 \quad \text{on} \quad r = 1, \tag{8.18}$$

the other conditions being unaltered.

Outer Solution

The parameters α, μ, and ε are all small, and so we can neglect the corresponding terms in (8.16), an approximation that should be valid away from the boundaries:

$$s \approx g \approx \ln \varepsilon + S, \tag{8.19}$$

where S satisfies

$$\nabla^2 S = c e^S, \tag{8.20}$$

with boundary conditions to be determined below. The reactant concentration satisfies

$$c_x = -\varepsilon \kappa c e^S, \tag{8.21}$$

and suggests that the relevant length scale for the reaction is $x \sim 1/\varepsilon\kappa$. Apparently (but see below) this is comparable to the tube length $1/\delta$ ($= 80$; $1/\varepsilon \approx 16$, $\kappa < 1$), and we thus suppose that c varies slowly with x, so that it can be taken as independent of x in solving for S. In particular, away from the inlet, S is determined by solving

$$S_{rr} + \frac{1}{r} S_r = c e^S. \tag{8.22}$$

Wall Boundary Layers

For $\varepsilon \ll 1$, there is a boundary layer near $r = 1$ where S becomes large. We put

$$r = 1 - \sqrt{\varepsilon} R, \tag{8.23}$$

so that to leading order (back with s)

$$S_{RR} \approx c e^s, \tag{8.24}$$

with

$$-\frac{\partial s}{\partial R} \approx \nu \approx \frac{(1 - \beta_1^4)\Lambda\sqrt{\varepsilon}}{\beta} \quad \text{on } R = 0. \tag{8.25}$$

Note that $\nu \approx 0.7$, so that $\nu = O(1)$. In order to match to the outer solution, we have to choose $s = s_0$ on $R = 0$, where

$$s_0 = \ln \frac{\nu^2}{2c}, \tag{8.26}$$

so that

$$s = -\ln\left(\frac{c}{2}\right) - 2\ln\left(R + \frac{2}{\nu}\right). \tag{8.27}$$

Writing this solution in terms of r and S implies that we require S in (8.22) to satisfy the matching condition

$$S \sim -\ln\frac{c}{2} - 2\ln(1 - r) \quad \text{as } r \to 1. \tag{8.28}$$

Suppose, for simplicity, that in fact $c \approx c(x)$. Then

$$S \approx \ln \frac{8}{c} - 2\ln(1 - r^2).$$ (8.29)

Thus, in the variables scaled by g_0^2/λ (≈ 70 K), the centre line (minimum) temperature is

$$s \approx \ln [8\varepsilon/c],$$ (8.30)

while the solid temperature at the wall is

$$s_0 \approx \ln \left[\frac{\varepsilon \Lambda^2 (1 - \beta_1^4)^2}{2c\beta^2} \right].$$ (8.31)

Gas Temperature Boundary Layer

Evidently, the assumption $\varepsilon \ll 1$ is hardly relevant to the above solution, since S can be found (for $c = c(x)$) explicitly. For the gas thermal boundary layer we put

$$r = 1 - \sqrt{\mu}\eta,$$ (8.32)

and if $\sqrt{\mu} \ll \sqrt{\varepsilon}$, then $s \approx s_0$ and to satisfy the approximate condition

$$-\frac{\gamma}{\Lambda\sqrt{\mu}} g_\eta \approx 1,$$ (8.33)

we have

$$g \approx s_0 + \frac{\Lambda\sqrt{\mu}}{\gamma} e^{-\eta},$$ (8.34)

while the gas temperature at the wall is

$$g_0 \approx s_0 + \frac{\Lambda\sqrt{\mu}}{\gamma}.$$ (8.35)

Inlet Boundary Layers

We omit details. There will be an adjustment layer for s of thickness $O(\sqrt{\varepsilon})$ and a thinner layer for g of thickness $\sqrt{\mu}$ (since $\sqrt{\mu} \gg \alpha$). In addition, (8.29) becomes valid (if $c = c(x)$) for distances $x \gg 1$, i.e. beyond the inlet region.

Temperature and Reactant Profiles

From (8.27) and (8.28), we can write a uniform approximation to the particle temperature as

$$s \approx \ln \frac{8\varepsilon}{c} - 2\ln\left[1 - r^2 + \frac{2\sqrt{\varepsilon}}{v}\right], \tag{8.36}$$

and the reactant concentration then satisfies (8.21), with s given as above. From this we have

$$c \approx 1 - \frac{8\varepsilon\kappa x}{\left(1 - r^2 + \frac{2\sqrt{\varepsilon}}{v}\right)^2}, \tag{8.37}$$

and the reaction zone boundary (where $c = 0$) is given by

$$x \approx \frac{\left(1 - r^2 + \frac{2\sqrt{\varepsilon}}{v}\right)^2}{8\varepsilon\kappa}, \tag{8.38}$$

and is shown in figure 8.2. The maximum extent of the reaction zone at the centre line is

$$x_m \approx \frac{\left(1 + \frac{2\sqrt{\varepsilon}}{v}\right)^2}{8\varepsilon\kappa}. \tag{8.39}$$

Similarly, a uniform expansion for the dimensionless gas temperature is

$$g \approx s + \frac{\Lambda\sqrt{\mu}}{\gamma} \exp\left[-\frac{(1-r)}{\sqrt{\mu}}\right]. \tag{8.40}$$

Figure 8.3 shows the cross-sectional profiles of the solid temperature given by (8.36) and (8.37) at various values of x. The core temperature varies slowly with x, while the solid warms up at the wall as the reaction proceeds. After the reaction front leaves the wall at $x \approx 1.4$, the core temperature begins to increase as the reactive core shrinks towards the centre. Figure 8.4 compares gas and solid temperatures given by (8.36), (8.37) and (8.40) at $x = 0.2$ near the tube inlet. The temperatures are the same in the core, but the gas has a hot boundary layer at the tube wall due to its relatively low thermal conductivity (as indicated by the low value of μ).

It is apparent that the solution for s becomes invalid if $c \to 0$ at the wall, since s must be bounded (by Λ) there, in contrast to (8.31), which indicates $s_0 \to \infty$ as $c \to 0$. At the wall, (8.37) implies that $c = c_0$ satisfies

$$c_0 \approx 1 - 2v^2\kappa x, \tag{8.41}$$

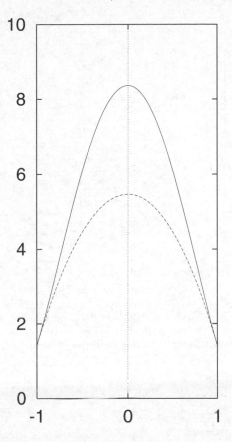

Fig. 8.2. The reaction zone given by the crude approximation (8.38) (solid curve), as well as the more accurate expression (8.52) (broken curve), with $\varepsilon = 0.06$, $\kappa = 0.72$, and $\nu = 0.7$

with a termination front at $x_0 = 1/2\nu^2\kappa$. Beyond the reaction front, we can expect a dead zone where $c = 0$, and in fact we see that the approximation for s is invalid for $x \gg 1$, both because (8.37) implies c depends on r, and also because of the dead zone. Despite this, we might expect the general characteristics of the reaction zone to be similar to that described here; that is to say, a central reactive core of inverted paraboloidal shape (as in figure 8.2) in which the gas and solid temperatures are depressed.

Reaction Front

The preceding analysis with $c = c(x)$ is inaccurate at large x but suggestive of the existence of a well-defined front at $r = r_f(x)$, say. If we now suppose

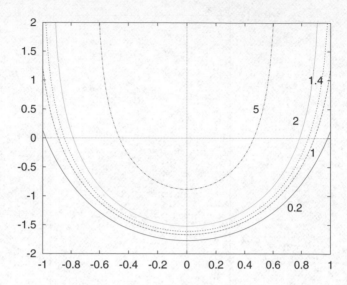

Fig. 8.3. Profiles of s versus r given by (8.36) and (8.37) for values of x chosen as 0.2, 1, 1.4, 2, 5, and with $\varepsilon = 0.06$, $\kappa = 0.72$, and $\nu = 0.7$. The reaction terminates on the wall at $x_0 \approx 1.417$, so that the profiles for $x = 2$ and $x = 5$ are not strictly accurate, though they do represent the qualitative behaviour of the solution

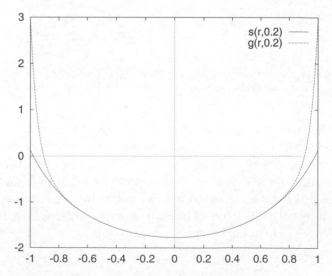

Fig. 8.4. A comparison of the gas and solid temperatures across the tube, at $x = 0.2$. The parameter values used are $\varepsilon = 0.06$, $\kappa = 0.72$, $\nu = 0.7$, $\Lambda = 6$, $\mu = 2.2 \times 10^{-3}$, and $\gamma = 0.1$

$c = c(x, r)$, then the problem we have to solve in $r < r_f$ for s and c is

$$\frac{c}{\varepsilon} e^s = S_{rr} + \frac{1}{r} s_r + s_{xx},$$

$$c_x = -\kappa c e^s, \qquad (8.42)$$

with $c = 1$ on $x = 0$, and

$$\frac{\partial s}{\partial r} \approx \frac{v}{\sqrt{\varepsilon}} \quad \text{on } r = 1. \qquad (8.43)$$

The solution with $c = c(x)$ is in fact valid so long as $\varepsilon \kappa x \ll 1$, and in particular while c_0 (on the wall, see (8.41)) is greater than zero, i.e. $x < x_0 = 1/2v^2\kappa$. For $x > x_0$, we suppose that $c = 0$ in $r > r_f$, so that

$$s = s_0 + \frac{v}{\sqrt{\varepsilon}} \ln r, \quad r > r_f, \qquad (8.44)$$

and the problem to determine r_f and s_0 is found from the solution for S ($= s - \ln \varepsilon$) of

$$c e^S = S_{rr} + \frac{1}{r} S_r,$$

$$c_x = -\varepsilon \kappa c e^S, \qquad (8.45)$$

with

$$c = 1 \quad \text{on} \quad x = 0, \quad c = 0 \quad \text{and} \quad S_r = \frac{v}{\sqrt{\varepsilon} r_f} \quad \text{on } r = r_f. \qquad (8.46)$$

It is straightforward to solve this problem numerically. Although

$$v/\sqrt{\varepsilon} \approx 3$$

is hardly large, it is useful to take the limit $\sqrt{\varepsilon} \ll 1$ seriously. In this case, a genuine reaction front exists. We revert to the variable s, and put

$$r = r_f - \sqrt{\varepsilon} R \qquad (8.47)$$

(cf. (8.23)), so that, to leading order,

$$s_{RR} \approx c e^s,$$

$$\rho c_R \approx c e^s, \qquad (8.48)$$

where we define

$$r_f' = -\kappa \sqrt{\varepsilon} \rho \ll 1. \qquad (8.49)$$

Thus the reaction zone will be of length $O(1/\kappa\sqrt{\varepsilon})$ in this case, and we can take $c \approx 1$ in $r < r_f$, so that suitable boundary conditions for matching purposes are

$$s \sim s^* - \frac{v}{r_f}R, \quad c \to 0 \quad \text{as} \quad R \to -\infty;$$

$$s_R \to 0, \quad c \to 1 \quad \text{as} \quad R \to +\infty : \tag{8.50}$$

here (cf. (8.44))

$$s^* = s_0 + \frac{v}{\sqrt{\varepsilon}}\ln r_f. \tag{8.51}$$

It follows that $\rho = v/r_f$, and thus from (8.49),

$$r_f^2 = 1 - 2\kappa\sqrt{\varepsilon}v(x - x_0), \tag{8.52}$$

where x_0 is given by $x_0 = 1/2v^2\kappa$ (see (8.41)), and also

$$c = 1 + \frac{s_R}{\rho}. \tag{8.53}$$

Thus s satisfies

$$s_{RR} \approx \left(1 + \frac{s_R}{\rho}\right)e^s, \tag{8.54}$$

with first integral

$$s_R - \rho\ln\left(1 + \frac{s_R}{\rho}\right) = \frac{1}{\rho}e^s; \tag{8.55}$$

the constant of integration has been chosen uniquely so that a monotone decreasing trajectory exists with $s_R \to 0$ as $R \to \infty$. The solution of (8.54) can be written parametrically as

$$s = 2\ln\rho + \ln[\xi - 1 + e^{-\xi}],$$

$$R = \frac{1}{\rho}\int_\xi^K \frac{du}{u - 1 + e^{-u}}, \tag{8.56}$$

where $\xi > 0$, and the value of K simply fixes the origin. For example, if we choose $c = \frac{1}{2}$ at $R = 0$, then $K = \ln 2$. The value of s_0 is then fixed via (8.50), together with the behaviour of (8.55) as $R \to -\infty$ ($\xi \to \infty$); we find

$$s_0 = \frac{v}{\sqrt{\varepsilon}}\ln\frac{1}{r_f} + \ln K + 2\ln\frac{v}{r_f} - \int_K^\infty \frac{(1 - e^{-u})du}{u(u - 1 + e^{-u})}, \tag{8.57}$$

so the wall temperature increases as the reacting core shrinks.

In $r < r_f$, $S = s - \ln \varepsilon$ then satisfies (with $c \approx 1$)

$$S_{rr} + \frac{1}{r} S_r = e^S \tag{8.58}$$

as before, with the matching condition obtained from (8.54),

$$S \sim \ln 2 - 2\ln(r_f - r) \quad \text{as} \quad r \to r_f, \tag{8.59}$$

with solution

$$S = \ln \left[\frac{8r_f^2}{(r_f^2 - r^2)^2} \right]. \tag{8.60}$$

8.4 Discussion

This basically concludes the analysis. We now have a good idea of how the solution behaves. There is a detailed reactive core of dimensional length (from (8.51)) $\approx (2\kappa v \sqrt{\varepsilon})^{-1} r_0$. Ignoring $\beta_1^4 \ll 1$, this is

$$x_c \approx \frac{\beta r_0}{2\kappa \Gamma} = \frac{u c_0 r_0 \Delta H}{2\varepsilon_G \sigma T^4}. \tag{8.61}$$

There is a preceding attached reactive core of length $x_a \approx (2v^2 \kappa)^{-1} r_0$; this is

$$x_a \approx \frac{\beta^2}{2\kappa \Lambda \Gamma} = \frac{u c_0 g_0^2 k_s \Delta H}{2\lambda \varepsilon_G^2 \sigma^2 T^8}. \tag{8.62}$$

In the reactive core, the temperature is low, and if the radiative heat flux is high enough ($v/\sqrt{\varepsilon} \gg 1$, i.e. $\Lambda \gg \beta$, i.e.

$$\frac{\lambda r_0 \varepsilon_G \sigma T^4}{g_0^2 k_s} \gg 1), \tag{8.63}$$

then the reaction occurs in a thin front.

There are three concluding statements to make concerning the success of this case study. Firstly, it is a good illustration of the efficacy of applied mathematics. In a matter of days, an experienced researcher can essentially solve this problem far beyond the ability of a numerical solution. Numerical solution of this problem is best provided as an illustration and confirmation of the analytic results.

Secondly, did this study answer the question *as posed* in the introduction? The answer is yes, in passing. One only has to nondimensionalise the equations to realise that the existence of thermal boundary layers at the walls and inlet will cause particular problems in devising numerical methods. This problem is best analysed first, and solved numerically later.

But the third and most significant observation is that all the analytic effort devoted to solving this problem in 1977 was directed towards solving the wrong problem. The model provided by BSC included the heat of reaction term $\mathcal{R}\Delta H$ in the *gas* temperature equation $((8.1)_1$, not $(8.1)_2)$. In addition the Arrhenius term was written in terms of the *gas* temperature, not the solid temperature. The consequent nature of the solution was quite different in its detail; for example $s = g$ was no longer appropriate.

Aris's two volumes [2] had appeared, as had Szekely et al.'s book [23]. They gave a comprehensive description of just this kind of problem. But the assembled applied analysts failed, in fact, to devote sufficient attention to the correct modelling of the problem, in their zeal to get on with its solution. If there is a lesson to be learned here, it is that one should never seek to solve fancy problems without a decent understanding of the appropriate physical processes.

8.5 Further Modelling Considerations

The reaction between a gas stream and a solid catalyst pellet is a surface reaction; it may occur at the pellet surface, or within the pellet if it is permeable. The simplest case is the "shrinking unreacted core" model, where the pellet is impermeable, and the heat of reaction is absorbed at the surface; see [23] or [8] for a discussion of this. It is then appropriate to include the heat absorption in the s equation, as we have done here, together with the macroscopic heat conduction term, since we can assume (since $\Gamma \sim 1$) that the pellets are in local thermal equilibrium.

The choice of the reaction rate $\mathcal{R}$ in (8.1) also omits any consideration of the surface nature of the reaction. In fact the conservation of gas reactant in $(8.1)_3$ would prescribe the supply from the gas stream to the particles, thus

$$u\frac{\partial c}{\partial x} = -h_c(c - c_s), \tag{8.64}$$

where h_c is a mass transfer coefficient, and c_s the pellet surface gas concentration. A first-order reaction would then prescribe the reaction rate as

$$\mathcal{R} = Kc_s e^{-\lambda/s}, \tag{8.65}$$

and it remains to determine c_s. (More generally c_s would be replaced by the fraction of adsorbed sites at the surface, but at low values of c_s, this is proportional to c_s, at least for the Langmuir isotherm.)

Finally, the rate of supply must equal the rate of reaction, so that

$$c_s = \frac{c}{1 + (K/h_c)e^{-\lambda/s}}. \tag{8.66}$$

In particular, we regain (8.1) if h_c is large enough; otherwise, the Arrhenius factor is modified.

Further understanding of how the surface reaction term is included requires a discussion of the detailed pellet dynamics. This is discussed in [23], and is lucidly reviewed in [17] and [7]. For the simplest model, the nonporous pellet alluded to above, the reaction causes the pellets to shrink. For complete gasification, or if any resulting ash is removed by spalling, the pellet size is given by the dimensionless radius ξ ($= 1$ initially), and this varies as the reaction proceeds. The volumetric rate of reaction then depends on ξ (see [8], pp. 195 ff.), and in a steady-state model the time evolution of ξ is described by an advective evolution equation which incorporates the solid's settling velocity.

The other type of pellet model describes porous pellets. When the pellet is formed from compacted grains, this is also referred to as a grain model. Following [8], the dimensionless temperature and concentration fields in a porous (compacted grain) pellet can be written in the form

$$\theta_t = \nabla^2\theta - \beta r,$$

$$c_t = \frac{1}{Le}\nabla^2 c - \mu r, \tag{8.67}$$

where the length scale is the pellet radius. If we realistically assume $\mu \ll 1$, then the pellet temperature θ is quasi-steady, and if also $\beta \ll 1$, then the pellets are isothermal. In this case, the evolution of the reaction in the pellet can be described through the reactant equations

$$\delta\frac{\partial c}{\partial \tau} = \frac{1}{\phi^2}\nabla^2 c - c\xi^{F_g-1}H(\xi),$$

$$\frac{\partial \xi}{\partial \tau} = -cH(\xi) \tag{8.68}$$

for a first-order reaction. The pellet temperature is normalised to zero here. In these equations, ξ is the (dimensionless) grain (not pellet) radius, $H(\xi)$ is the Heaviside step function, $\delta = c_s/\rho_s$ is the ratio of gas molar concentration to solid molar concentration, $\tau = \delta\mu t$, and $\phi = \{\mu Le\}^{1/2}$ is the Thiele modulus.

The exponent F_g represents grain shape, being equal to 1, 2, or 3 for plates, cylinders, or spheres respectively.

Realistically, $\delta \ll 1$, and the time derivative of c can be ignored (the quasi-static assumption). In general, solutions must be numerical, but analytic results are possible for large or small Thiele modulus. If $\phi \ll 1$ (chemical control), the reactant concentration is uniform in the pellet, and reaction proceeds uniformly throughout. More interesting mathematically is the diffusion-controlled limit $\phi \gg 1$, when a reaction front moves into the pellet (see below).

Many other aspects of the modelling may be considered as exercises. For example:

- Why does the fact that $\Gamma \sim 1$ imply that we can assume that pellets are in thermal quasi-equilibrium?
- Follow the discussion in [8] to modify the model (8.1) to allow for shrinking nonporous pellets, and examine the effect of this modification on the analysis presented here.
- Use the discussion in [8] (pp. 190 ff.) and [23] (pp. 133 ff.) to derive (8.68). For the quasi-static chemically controlled pellet, derive the appropriate form of the heat sink term in (8.1).
- Show that for the quasi-static diffusionally controlled pellet with spherical grains ($F_g = 3$), a reaction front at $r = r_f$, which separates a reacted fringe ($\xi = 0$) from an unreacted core ($c \approx 0, \xi \approx 1$), moves into the pellet at a rate

$$\dot{r}_f \approx -\frac{3}{\phi^2 r_f (1 - r_f)}.$$

Deduce the corresponding form of the heat sink term in (8.1).

References

[1] Adrover, A. & Giona, M. (1997) Solution of unsteady-state shrinking-core models by means of spectral/fixed-point methods: nonuniform reactant distribution and nonlinear kinetics. Ind. Eng. Chem. Res. **36**, 2452–2465.

[2] Aris, R. (1975) *Mathematical Theory of Diffusion and Reaction in Permeable Catalysts*, (2 volumes), Oxford University Press.

[3] Bhatia, S. K. (1985) On the pseudo-steady state hypothesis for fluid solid reactions. Chem. Eng. Sci. **40**, 868–872.

[4] Bhatia, S. K. (1991) Perturbation analysis of gas-solid reactions. 2. Reduction to the diffusion-controlled shrinking core. Chem. Eng. Sci. **46**, 1465–1474.

[5] Cao, G., Varma, A. & Strieder, W. (1993) Approximate solutions for nonlinear gas-solid noncatalytic reactions. AIChE J. **39**, 913–917.

[6] Chan, Y. H. & McElwain, D. L. S. (1996) Asymptotic analysis and effect of reaction order on a reversible gas-solid system with fast reaction. J. Eng. Math. **30**, 365–386.

[7] Doraiswamy, L. K. & Sharma, M. M. (1984) *Heterogeneous Reactions: Analysis, Examples and Reactor design, Vol. 1*. John Wiley, New York, USA.

[8] Fowler, A. C. (1997) *Mathematical Models in the Applied Sciences*, Cambridge University Press, England.

[9] Hahn, Y. B. & Chang, K. S. (1998) Mathematical modelling of the reduction process of iron ore particles in two stages of twin-fluidized beds connected in series. Metall. Mater. Trans. B **29**, 1107–1115.

[10] Jamshidi, E. & Ebrahim, H. A. (1996) An incremental analytical solution for gas-solid reactions, application to the grain model. Chem. Eng. Sci. **51**, 4253–4257.

[11] Lédé, J. & Villermaux, J. (1993) Comportement thermique et chimique de particules solides subissant une réaction de décomposition endothermique sous l'action d'un flux de chaleur externe. Can. J. Chem. Eng. **71**, 209–217.

[12] Liu, F., McElwain, D. L. S. & Donskoi, E. (1998) The use of a modified Petrov-Galerkin method for gas-solid reaction modelling. IMA J. Appl. Math. **61**, 33–46.

[13] Lu, H. B., Mazet, N., Coudevylle, O. & Mauran, S. (1997) Comparison of a general model with a simplified approach for the transformation of solid-gas media used in chemical heat transformers. Chem. Eng. Sci. **52**, 311–327.

[14] Lu, H. B., Mazet, N. & Spinner, B. (1996) Modeling of gas-solid reaction—coupling of heat and mass transfer with chemical reaction. Chem. Eng. Sci. **51**, 3829–3845.

[15] Patisson, F., Francois, M. G. & Ablitzer, D. (1998) A non-isothermal, non-equimolar transient kinetic model for gas-solid reactions. Chem. Eng. Sci. **53**, 697–708.

[16] Pritsker, M. D. (1996) Shrinking core model for systems with facile heterogeneous and homogeneous reactions. Chem. Eng. Sci. **51**, 3631–3645.

[17] Ramachandran, P. A. & Doraiswamy, L. K. (1982) Modeling of non-catalytic gas-solid reactions. AIChE J. **28**, 881–900.

[18] Rode, H., Orlicki, D. & Hlavacek, V. (1995) Reaction-rate modelling in noncatalytic gas-solid systems—species transport and mechanical stress. AIChE J. **41**, 2614–2624.

[19] Shah, N. & Ottino, J. M. (1987) Transport and reaction in evolving, disordered composites. 1. Gasification of porous solids. Chem. Eng. Sci. **42**, 63–72.

[20] Shankar, K. & Yortsos, Y. C. (1983) Asymptotic analysis of single pore gas-solid reactions. Chem. Eng. Sci. **38**, 1159–1165.

[21] Sohn, H. Y. & Chaubal, P.C. (1986) Approximate closed-form solutions to various model equations for fluid-solid reactions. AIChE J. **32**, 1574–1578.

[22] Sohn, H. Y., Johnson, S. H. & Hindmarsh, A. C. (1985) Application of the method of lines to the analysis of single fluid-solid reactions in porous solids. Chem. Eng. Sci. **40**, 2185–2190.

[23] Szekely, J., Evans, J. W. & Sohn, H. Y. (1976) *Gas-Solid Reactions*, Academic Press, New York, USA.

[24] Taylor, P. R., Dematos, M. & Martins, G. P. (1983) Modeling of non-catalytic fluid-solid reactions—the quasi-steady state assumption. Metall. Trans. B **14**, 49–53.

[25] Zhou, L. & Sohn, II. Y. (1996) Mathematical modelling of fluidized-bed chlorination of rutile. AIChE J. **42**, 3102–3112.

Andrew C. Fowler
OCIAM
Mathematical Institute, 24–29, St. Giles, Oxford OX1 3LB, UK
fowler@maths.ox.ac.uk

9

Simulation of the Temperature Behaviour

of Hot Glass during Cooling

Preface

The Institute for Industrial Mathematics, which will become the first Fraunhofer Institute for applied mathematics in Germany, has focused its work on practical problems coming from industry. The following case study deals with a very old problem in the production of glass. It is part of a project of the ITWM with Schott Glas, the most important producer of specialized glasses in Germany. The famous physicist Josef von Fraunhofer faced the problem when trying to produce large lenses for astronomical binoculars – almost 200 years ago. During the cooling of the glass, thermal tensions tended to produce fractures, which might be avoided by a proper control of the cooling process. This can be achieved by regulation of the surrounding temperature. However, in order to control the process, it is necessary to understand and to predict the thermal tensions, i.e. the stresses set up by nonhomogeneous thermal contraction. This is complicated, because heat is transported through radiation. Glass is a semitransparent medium – each point in the glass is a new source of radiation. This property is modelled in the radiation equation – but this equation is not easy to solve. In contrast to other case studies in this volume, there is in general nothing flat or thin. The thermal tensions depend heavily on the three-dimensional object as a whole, and the object very often has no symmetries. Therefore, numerical methods are unavoidable, but still pose difficult problems: a straightforward solution of the three-dimensional radiative equation, say by ray tracing, discrete ordinates, or even P_1-approximation does not work in real (i.e. real industrial) problems. Schott asked us to develop a numerical method for the radiative transfer which is

- as fast as possible
- as accurate as necessary
- able to be easily implementable into commercial software packages.

A way to solve this problem is discussed in the following case study, where we will employ asymptotic analysis to construct practical approximation methods.

9.1 Cooling of Glass

Knowledge of the temperature distribution during the cooling process of glass is very important to control the internal stress field and finally the quality of the glass. Heat transfer in glass as a semitransparent material is accomplished not only by conduction but also by radiation. This is particularly the case at the beginning of the annealing process when the temperature of glass is higher than 400°C. From numerical experiments (for instance in the one-dimensional case), it is known that, at this temperature, radiative transfer dominates conduction.

In a semitransparent material, radiation is absorbed and emitted at each point in the interior. The average distance a photon travels before interacting depends on the wavelength of the radiation. The part of the electromagnetic spectrum for which the mean free path of the photons is much smaller than the dimension of the material considered is called the opaque region. There, all radiation is absorbed immediately. Only at the surface of the glass can radiation be emitted from the opaque spectrum into the surroundings. In the opposite case, the transparent region, where the mean free path of photons is much bigger than the glass dimension, does not give any contribution to the energy exchange. Photons of the so-called semitransparent wavelength region are emitted as well as absorbed inside the glass.

In the following we concentrate on mathematical approaches for three-dimensional radiative heat transfer. For the simulation of the temperature behaviour of hot glass during cooling, a solution procedure is required, which is as accurate as necessary and as fast as possible.

9.2 Mathematical Formulation of the Problem

9.2.1 Heat and Radiative Transfer Equations

We consider an open bounded three-dimensional domain G with an initial temperature distribution $T_0(r)$. The transfer of energy can be described by the heat transfer equation, which can be found in standard books of mathematical physics (e.g.[12]):

$$c_m \rho_m \frac{\partial T}{\partial t}(r, t) = -\nabla \cdot q(r, t), \ r \in \bar{G}, \ t \in [0, t^\star]$$

$$T(r, 0) = T_0(r), \ r \in \bar{G},$$

(9.1)

where c_m is the specific heat, ρ_m the density, T is the temperature, r the position vector, t is the time, and q the heat flux vector. The heat flux vector in this case consists of two parts: q_c – the result of heat conduction and q_r – the result of radiation. For the conductive part, we assume that Fourier's law is valid so that

$$q_c(r, t) = -k_h \nabla T(r, t),$$

where k_h denotes the thermal conductivity. For determining the radiative flux vector q_r we refer to [11] or [14]. There we find that q_r is defined as the first moment of the spectral radiative intensity $I(r, \Omega, \lambda)$ with respect to the three-dimensional directional vector Ω:

$$q_r = \int_0^\infty \int_{S^2} I(r, \Omega, \lambda) \Omega \, d\Omega \, d\lambda, \tag{9.2}$$

where S^2 denotes the surface of the unit sphere. The spectral radiative intensity $I(r, \Omega, \lambda)$ is a function of the position $r \in S^2$, the direction of the ray Ω, and the wavelength λ. It describes the radiative energy flow per unit solid angle, per unit area normal to the ray direction, and per wavelength.

$I(r, \Omega, \lambda)$ follows from the radiative transfer equation

$$\Omega \cdot \nabla I(r, \Omega, \lambda) + \kappa(\lambda) I(r, \Omega, \lambda) = \kappa(\lambda) B(T(r), \lambda)$$
$$(r, \Omega) \in (\bar{G} \times S^2) \setminus (\partial G \times S^2|_{n \cdot \Omega < 0}), \tag{9.3}$$

which describes an attenuation proportional to the absorption coefficient $\kappa(\lambda)$, depending on the wavelength, and an augmentation by emission related to the Planck's function [10]:

$$B(T(r), \lambda) = \frac{2hc_0^2}{n_g^2 \lambda^5 (e^{hc_0/n_g \lambda kT} - 1)}.$$

Here c_0 denotes the speed of light, h Planck's constant, n_g the refractive index of glass, and k Boltzmann's constant.

The absorption coefficient was given by the industrial partner in the form of a piecewise constant function:

$$\kappa(\lambda) = \kappa_k, \quad \lambda \in [\lambda_k, \lambda_{k+1}], \quad k = 1, 2, \dots M_k, \tag{9.4}$$

where $\kappa_k \in [0, \infty]$, i.e. $\kappa_k = \infty$ is possible. We call these cases, for which $\kappa_k = \infty$, "opaque".

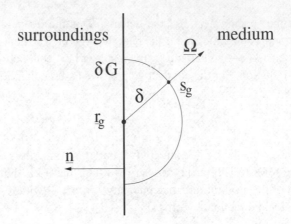

Fig. 9.1. Boundary notation

Using the assumption (9.4) and the notation

$$I^k(r, \mathbf{\Omega}) = \int\limits_{\lambda_k}^{\lambda_{k+1}} I(r, \mathbf{\Omega}, \lambda)d\lambda, \quad B^k(T(r)) = \int\limits_{\lambda_k}^{\lambda_{k+1}} B(T(r), \lambda)d\lambda,$$

the radiative transfer equation can be written in the form

$$\mathbf{\Omega} \cdot \nabla I^k(r, \mathbf{\Omega}) + \kappa_k I^k(r, \mathbf{\Omega}) = \kappa_k B^k(T(r)). \qquad (9.5)$$

To solve this equation, one has to add appropriate boundary conditions. In our case the surface of the glass is covered with soot. Therefore we have black surfaces. The incoming radiation at the boundary ∂G of our domain is given by the black body intensity

$$I^k(r_g, \mathbf{\Omega}) = B^k(T_a(r_g)), \quad r_g \in \partial G, \quad n \cdot \mathbf{\Omega} < 0, \qquad (9.6)$$

where T_a is the temperature of the surroundings.

9.2.2 Modelling of the Boundary Conditions for the Heat Transfer Equation

We consider a small hemispherical neighbourhood, with radius δ, of $r_g \in \partial G$ (see figure 9.1) and define the difference of the radiative flux through the boundary point r_g and the inner point s_g, which is placed at a distance δ from r_g in direction $\mathbf{\Omega}$.

In the following we determine the difference between the normal fluxes through the points r_g and s_g. We have

$$n \cdot q(r_g) - n \cdot q(s_g) = \int_0^\infty \left(\int_{n \cdot \Omega > 0} n \cdot \Omega [I(r_g, \Omega, \lambda) - I(s_g, \Omega, \lambda)] d\Omega \right) d\lambda$$

$$+ \int_0^\infty \left(\int_{n \cdot \Omega < 0} n \cdot \Omega [I(r_g, \Omega, \lambda) - I(s_g, \Omega, \lambda)] d\Omega \right) d\lambda$$

$$(9.7)$$

First we consider the semitransparent case: $\kappa(\lambda) < \infty$. From the radiative transfer equation (9.3) we obtain the formal solution as

$$I(r, \Omega, \lambda) = I(r_g, \Omega, \lambda) e^{-\kappa(\lambda) d(r, \Omega)}$$

$$+ \kappa(\lambda) \int_0^{d(r, \Omega)} B(T(r_g + s\Omega), \lambda) e^{-\kappa(\lambda)(d(r, \Omega) - s)} ds, \quad (9.8)$$

where $d(r, \Omega)$ is the distance between the points r and r_g in the direction $-\Omega$.

- $n \cdot \Omega > 0$: In this case, the radiative intensity coming from the direction Ω does not originate from the boundary. Therefore the intensity at both points is given by (9.8). The limit $\delta \to 0$ leads to

$$I(r_g, \Omega, \lambda) = I(s_g, \Omega, \lambda).$$

- $n \cdot \Omega < 0$: From (9.8) we have

$$I(s_g, \Omega, \lambda) = I(r_g, \Omega, \lambda) e^{-\kappa(\lambda)\delta}$$

$$+ \kappa(\lambda) \int_0^\delta B(T(r_g + s\Omega), \lambda) e^{-\kappa(\lambda)(\delta - s)} ds.$$

In the limit $\delta \to 0$ we have:

$$I(r_g, \Omega, \lambda) = I(s_g, \Omega, \lambda).$$

Now we consider the opaque case: $\kappa(\lambda) = \infty$. From the radiative transfer equation it follows that

$$I(r, \Omega, \lambda) = B(T(r), \lambda). \quad (9.9)$$

- $n \cdot \Omega > 0$: Now the radiation goes from the inner point s_g to the boundary point r_g. Because of the continuity of the black body function B, it follows that, as $\delta \to 0$,

$$I(r_g, \Omega, \lambda) = I(s_g, \Omega, \lambda).$$

- $n \cdot \Omega < 0$: Here the radiation goes from the boundary point r_g to the inside of the medium s_g. From (9.9) we get, for $\delta \to 0$,

$$I(r_g, \Omega, \lambda) - I(s_g, \Omega, \lambda) = B(T_a(r), \lambda) - B(T(r), \lambda).$$

Since this is the only case for which $I(r_g, \Omega, \lambda) \neq I(s_g, \Omega, \lambda)$, this will provide the only contribution to the right-hand side of (9.7).

Hence we obtain for the radiative flux at the boundary, from (9.7),

$$q(r_g) = \int_{\Lambda_O} [B(T_a(r), \lambda) - B(T(r), \lambda)]d\lambda \int_{n \cdot \Omega < 0} n \cdot \Omega d\Omega$$

$$= \pi \int_{\Lambda_O} [B(T_a(r), \lambda) - B(T(r), \lambda)]d\lambda \qquad (9.10)$$

where Λ_O denotes the set of all wavelengths which are opaque, i.e.

$$\Lambda_O := \{\lambda | \kappa(\lambda) = \infty\} \qquad (9.11)$$

We therefore obtain the following boundary condition for the heat transfer equation:

$$k_h \frac{\partial T}{\partial n}(r_g, t) = \pi \int_{\Lambda_O} [B(T_a(r), \lambda) - B(T(r), \lambda)]d\lambda. \qquad (9.12)$$

The foregoing analysis is summarized in the following sequence of tasks. Solve the radiative transfer equations (9.5)–(9.6) for every spectral band k to get $I^k(r, \Omega)$. Use the definition (9.2) to calculate the divergence of the radiative flux vector as

$$\nabla \cdot q(r) = \sum_{k=1}^{M_k} \int_{S^2} I^k(r, \Omega)d\Omega.$$

Finally, solve the heat transfer equation (9.1) with the boundary conditions (9.12).

9.3 Numerical Solution Methods

9.3.1 The Heat Transfer Equation

The numerical solution of the heat transfer equation is a standard problem. For instance, one can use the finite volume method which guarantees us an energy-preserving discretization. To do this, we make a discretization of the time interval: $0 = t_0 < t_1 < \cdots < t_L = t^\star$. We divide the three-dimensional domain G into N finite volume elements G_j (with N grid points r_j) and integrate equation (1) over each G_j and time interval $[t_l, t_{l+1}]$. Using the Gauss formulae, one can transform these volume integrals into integrals over the surfaces of G_j. Approximating the first-order derivatives by finite differences, one obtains an algebraic equation for each volume integral. Finally, we have to solve a system of algebraic equations in every time step. More about the finite discretization can be found in detail in many books about numerical methods for PDEs (for instance [1]).

9.3.2 Ray Tracing

Now, we focus our attention on the solution of the radiative transfer equation. One of the simplest methods for solving the radiative transfer equation is a ray-tracing technique. This method is very popular for instance in computer graphics. There, one follows the rays from the eye of the observer into the scene within a certain solid angle. Using ray tracing for radiative transfer is more complex because every spatial point r in G is a new source, and for every space point one has to trace back the rays in all directions of the three-dimensional domain.

So, let r be a grid point coming from the discretization of the domain G for the heat transfer equation. Next, we discretize the total solid angle of 4π by M components:

$$\mathbf{\Omega}_1, \mathbf{\Omega}_2, \dots, \mathbf{\Omega}_M.$$

For every grid point r and every discrete direction $\mathbf{\Omega}_i$, we follow the ray back to the boundary. The radiative transfer equation can be written as

$$-\frac{d}{ds}[I^k(r - s\mathbf{\Omega}_i, \mathbf{\Omega}_i)] = \kappa_k(B^k(T(r - s\mathbf{\Omega}_i)) - I^k(r - s\mathbf{\Omega}_i, \mathbf{\Omega}_i)), \quad (9.13)$$

$$I^k(r_g, \mathbf{\Omega}_i) = B(T(r_g)).$$

Thus, we have to solve an ordinary differential equation along the path $r - s\mathbf{\Omega}_i$ (see figure 9.2).

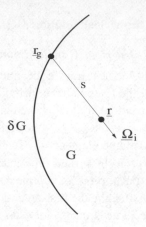

Fig. 9.2. Ray tracing

For the solution of (9.14) one can use different numerical methods such as the Euler method or Runge–Kutta methods. We take the formal solution of (9.14):

$$I^k(r, \Omega_i) = B^k(T_a(r_g))e^{-\kappa_k d(r, \Omega_i)}$$

$$+ \kappa_k \int_0^{d(r, \Omega_i)} B^k(T(r_g + s\Omega_i))e^{-\kappa_k(d(r, \Omega_i) - s)} ds. \qquad (9.14)$$

Let $r_g = x_1, x_2, \ldots, x_l, x_{l+1}, \ldots, x_L = r$ be the spatial discretization of the path, given by s_i, $i = 1, \ldots, L$. We consider a linear approximation of $B^k(T)$ between the points x_l and x_{l+1}. Then the formal solution on this interval can be approximated as

$$I^k(x_{l+1}, \Omega_i) = I^k(x_l, \Omega_i)e^{-\kappa_k h_l} + B^k(T(x_{l+1})) - B^k(T(x_l))e^{-\kappa_k h_l}$$

$$- \frac{B^k(T(x_{l+1})) - B^k(T(x_l))}{\kappa_k h_l}(1 - e^{-\kappa_k h_l}),$$

where

$$h_l = |x_{l+1} - x_l|.$$

From the construction of the ray-tracing method, it is clear that the set $\{r_j\}$ which contains the discretization points for the heat transfer equation and $\{x_l\}$ which denotes the set of discretization points along the ray are in general different. Therefore, one has to interpolate $B^k(T(x_l))$ from the values of $B^k(T(r_j))$.

The heat transfer equation may be solved using a finite volume method. The temperature for every finite volume is known from the previous time step. To determine the temperature for x_l it is sufficient to determine the finite volume element to which this grid point belongs.

The next question is the selection of the discrete ordinates $\mathbf{\Omega}_i$. Here, we refer to the literature, for instance [3], [4], [8]. We are now able to solve the radiative transfer equation. The results of the simulation are shown in the next chapter.

It turns out that the ray-tracing technique we have described is expensive in time and memory. This explains why it is not satisfactory for an industrial client. For industrial applications, a faster method is needed which allows the use of computers of the workstation class. Our approach leads to the diffusion approximation.

9.3.3 A Diffusion Approximation

One of the first diffusion approximations was developed by S. Rosseland at the beginning of the twentieth century. The idea of the method is to express the radiative flux in the same manner as the heat flux:

$$q_r = -k_r(T)\nabla T(r, t). \tag{9.15}$$

We know that conduction is a local phenomenon, whereas radiation can be a local as well as a global phenomenon. The character of radiation depends on the magnitude of the absorption coefficient κ_k. To get a local description of the radiative flux like (9.15), we assume that the absorption coefficient is large. Then the radiative transfer equation can be rewritten as

$$\frac{1}{\kappa_k}\mathbf{\Omega} \cdot \nabla I^k(r, \mathbf{\Omega}) + I^k(r, \mathbf{\Omega}) = B^k(T(r)), \tag{9.16}$$

where $1/\kappa_k$ is a small parameter. Therefore one may use a simple perturbation analysis to solve (9.16).

We expand the spectral radiative intensity as a series in the small parameter:

$$I^k(r, \mathbf{\Omega}) = I_0^k(r, \mathbf{\Omega}) + \frac{1}{\kappa_k}I_1^k(r, \mathbf{\Omega}) + \cdots. \tag{9.17}$$

Substituting this expansion into (9.16) and collecting terms of the same order, we obtain

$$\left(\frac{1}{\kappa_k}\right)^0: \qquad I_0^k(r, \mathbf{\Omega}) = B^k(T(r)),$$

$$\left(\frac{1}{\kappa_k}\right)^1 : \quad I_1^k(\mathbf{r}, \mathbf{\Omega}) = -\mathbf{\Omega} \cdot \nabla B^k(T(\mathbf{r})),$$

$$\ldots \qquad\qquad \ldots$$

As a first-order approximation for the spectral radiative intensity, we obtain

$$I^k(\mathbf{r}, \mathbf{\Omega}) \approx B^k(T(\mathbf{r})) - \frac{1}{\kappa_k}\mathbf{\Omega} \cdot \nabla B^k(T(\mathbf{r})). \tag{9.18}$$

Taking the first moment of (9.18) with respect to $\mathbf{\Omega}$, we finally arrive at the desired expression for the radiative flux vector:

$$\mathbf{q}_r(\mathbf{r}) = -\frac{4\pi}{3}\left(\sum_{k=1}^{M_k}\frac{1}{\kappa_k}\frac{dB^k}{dT}(T(\mathbf{r}))\right)\nabla T(\mathbf{r}). \tag{9.19}$$

The resulting diffusion equation is known as the Rosseland approximation.

Following the procedure of section 9.2.2, one may find the appropriate boundary condition for the Rosseland approximation, which is known in the literature as the Deissler jump condition. Thus, instead of problem (9.1) with (9.12), we have to solve the problem

$(\mathbf{r} \in G, t \in [0, t^\star])$:

$$c_m\rho_m\frac{\partial T}{\partial t}(\mathbf{r}, t) = \nabla \cdot \left(\left(k_h + \frac{4\pi}{3}\sum_{k=1}^{M_k}\frac{1}{\kappa_k}\frac{dB^k}{dT}(T(\mathbf{r}))\right)\nabla T(\mathbf{r})\right),$$

$(\mathbf{r} \in G)$: $T(\mathbf{r}, 0) = T_0(\mathbf{r}),$

$(\mathbf{r}_g \in \partial G)$:

$$\left(k_h + \frac{2\pi}{3}\sum_{k=1}^{M_k}\frac{1}{\kappa_k}\frac{dB^k}{dT}(T(\mathbf{r}_g))\right)\frac{\partial T}{\partial n}(\mathbf{r}_g, t) = \sigma(T_a^4(\mathbf{r}_g, t) - T^4(\mathbf{r}_g, t)),$$

where σ is the Stefan–Boltzmann constant.

The simulation of the temperature distribution using the Rosseland approximation does not give sufficiently accurate results. It is worthwhile pointing out that the Rosseland approximation is the standard method for radiative heat transfer in the glass industry.

9.3.4 Two-Scale Analysis

Because of the insufficiently accurate results of the Rosseland approximation, other approaches must be explored, and we propose a two-scale analysis.

The Rosseland approximation as a "one-scale" asymptotic is a sufficient tool to resolve the intensity in the bulk region, where the temperature varies only on the length scale of the geometry. But it is not able to correctly resolve the intensity close to the boundary, where the influence of the boundary changes on a smaller length scale due to the exponential decay of intensity coming from the boundary. The Deissler jump condition overcomes this problem only for very large absorption coefficients κ_k, where the influence of the boundary can be reduced to the boundary itself. Therefore the Rosseland approximation is only applicable for very large absorption coefficients. A boundary layer seems to be indicated to solve the radiative transfer even for lower absorption coefficients, but often this layer has to cover the whole glass, so that nothing is gained. This is the reason why we introduce two space coordinates corresponding to two different length scales. The first coordinate resolves the temperature variation which lives on a length scale 1, and the second coordinate resolves the boundary influence which varies on a length scale $1/\kappa_k$, the mean path of rays. Following this approach, we obtain a two-scale asymptotic expansion for the radiative heat transfer equation, which acts as a correction of the heat diffusion equation.

We introduce a two-scale radiative intensity $I^k(r, s, \Omega)$ which depends on the original space coordinate r, the rescaled space coordinate of the mean free path of rays $s = \kappa_k r$, and the directional vector Ω. The original radiative intensity is related to the two-scale radiative intensity as follows.

$$I^k(r, \Omega) = I^k(r, \kappa_k r, \Omega). \tag{9.20}$$

The radiative transfer equation for this two-scale radiative intensity then reads

$$\frac{1}{\kappa_k} \Omega \cdot \nabla_r I^k(r, s, \Omega) + \Omega \cdot \nabla_s I^k(r, s, \Omega) + I^k(r, s, \Omega) = B^k(T(r)). \tag{9.21}$$

The boundary condition for the two-scale radiative intensity is not completely prescribed. The only thing which is prescribed is that the original radiative intensity has to fulfil the original boundary condition, i.e.

$$I^k(r_g, \kappa_k r_g, \Omega) = B^k(T_a), \qquad r_g \in \partial G, \ n \cdot \Omega < 0. \tag{9.22}$$

We now assume an asymptotic expansion for the solution. Assuming that the absorption coefficient κ_k is large, we expand the radiative intensity in a series in powers of $1/\kappa_k$, i.e.

$$I^k(r, s, \Omega) = I_0^k(r, s, \Omega) + \frac{1}{\kappa_k} I_1^k(r, s, \Omega) + \cdots. \tag{9.23}$$

Substituting this expansion in the two-scale radiative transfer equation (9.21) and sorting terms of same order, we obtain

$$\mathbf{\Omega} \cdot \nabla_s I_0^k(r, s, \mathbf{\Omega}) + I_0^k(r, s, \mathbf{\Omega})$$

$$+ \frac{1}{\kappa_k} \left(\mathbf{\Omega} \cdot \nabla_r I_0^k(r, s, \mathbf{\Omega}) + \mathbf{\Omega} \cdot \nabla_s I_1^k(r, s, \mathbf{\Omega}) + I_1^k(r, s, \mathbf{\Omega}) \right) + \cdots$$

$$= B^k(T(r)). \tag{9.24}$$

In order to obtain convergence for the series (9.23) for $\kappa_k \to \infty$, the terms of different order in (9.24) have to be satisfied separately.

9.3.4.1 Zeroth-Order Expansion

Collecting the terms of order κ_k^{0} we obtain the equation

$$\mathbf{\Omega} \cdot \nabla_s I_0^k(r, s, \mathbf{\Omega}) + I_0^k(r, s, \mathbf{\Omega}) = B^k(T(r)) \tag{9.25}$$

in conjunction with the boundary condition

$$I_0^k \left(\frac{1}{\kappa_k} s_g, s_g, \mathbf{\Omega} \right) = B^k(T_a), \qquad s_g \in \partial(\kappa_k G), \; n \cdot \mathbf{\Omega} < 0. \tag{9.26}$$

As already mentioned, the boundary condition is not completely prescribed. It only gives a boundary condition for $r_g = (1/\kappa_k)s_g$. However, we can extend it to a boundary condition for all $r \in \bar{G}$, i.e.

$$I_0^k(r, s_g, \mathbf{\Omega}) = B^k(T_a), \qquad r \in G, \; s_g \in \partial(\kappa_k G), \; n \cdot \mathbf{\Omega} < 0. \tag{9.27}$$

Equation (9.25) with the boundary condition (9.27) is a differential equation only in s and $\mathbf{\Omega}$, whereas the right-hand side depends only on r. Due to the linearity of the equation, it is sufficient to solve the equation for the right-hand side equal to 1. This solution will be denoted by $I^k[1]$, i.e.

$$\mathbf{\Omega} \cdot \nabla_s I^k[1](s, \mathbf{\Omega}) + I^k[1](s, \mathbf{\Omega}) = 1, \tag{9.28}$$

$$I^k[1](s_g, \mathbf{\Omega}) = 0, \qquad s_g \in \partial(\kappa_k G), \; n \cdot \mathbf{\Omega} < 0. \tag{9.29}$$

The zeroth-order expansion can then be expressed as

$$I_0^k(r, s, \mathbf{\Omega}) = B^k(T_a) + \left(B^k(T(r)) - B^k(T_a) \right) I^k[1](s, \mathbf{\Omega}). \tag{9.30}$$

9.3.4.2 First-Order Expansion

Collecting the terms of order κ_k^{-1}, we obtain

$$\mathbf{\Omega} \cdot \nabla_s I_1^k(r, s, \mathbf{\Omega}) + I_1^k(r, s, \mathbf{\Omega}) = -\mathbf{\Omega} \cdot \nabla_r I_0^k(r, s, \mathbf{\Omega}), \tag{9.31}$$

and by making use of the solution for I_0^k, we have

$$\mathbf{\Omega} \cdot \nabla_s I_1^k(\mathbf{r}, s, \mathbf{\Omega}) + I_1^k(\mathbf{r}, s, \mathbf{\Omega}) = -\sum_{i=1}^{3} \frac{\partial B^k}{\partial r_i}(T(\mathbf{r})) \, \Omega_i I^k[1](s, \mathbf{\Omega}).$$

(9.32)

Additionally, we obtain the boundary condition

$$I_1^k(\mathbf{r}, s_g, \mathbf{\Omega}) = 0, \qquad \mathbf{r} \in \bar{G}, \ s_g \in \partial(\kappa_k G), \ \mathbf{n} \cdot \mathbf{\Omega} < 0. \qquad (9.33)$$

Equation (9.32) with the boundary condition (9.27) is again only a linear differential equation in s and $\mathbf{\Omega}$, such that factors appearing on the right-hand side and depending only on $\mathbf{r}$ can be extracted. Therefore it remains to solve the equations,

$$\mathbf{\Omega} \cdot \nabla_s J_i^k(s, \mathbf{\Omega}) + J_i^k(s, \mathbf{\Omega}) = -\Omega_i I^k[1](s, \mathbf{\Omega}), \qquad i = 1, 2, 3, \qquad (9.34)$$

in conjunction with the boundary conditions

$$J_i^k(s_g, \mathbf{\Omega}) = 0, \qquad s_g \in \partial(\kappa_k G), \ \mathbf{n} \cdot \mathbf{\Omega} < 0. \qquad (9.35)$$

It is easy to verify that

$$J_i^k(s, \mathbf{\Omega}) = -s_i I^k[1](s, \mathbf{\Omega}) + I^k[s_i](s, \mathbf{\Omega}), \qquad i = 1, 2, 3, \qquad (9.36)$$

are solutions to these differential equations, where $I^k[1]$ denotes again the solution of (9.28) and (9.29), and the $I^k[s_i]$ denote the solutions of the differential equations

$$\mathbf{\Omega} \cdot \nabla_s I^k[s_i](s, \mathbf{\Omega}) + I^k[s_i](s, \mathbf{\Omega}) = s_i, \qquad i = 1, 2, 3, \qquad (9.37)$$

$$I^k[s_i](s_g, \mathbf{\Omega}) = 0, \qquad s_g \in \partial(\kappa_k G), \ \mathbf{n} \cdot \mathbf{\Omega} < 0. \qquad (9.38)$$

The first-order expansion can then be expressed as

$$I_1^k(\mathbf{r}, s, \mathbf{\Omega}) = \sum_{i=1}^{3} \frac{\partial B^k}{\partial r_i}(T(\mathbf{r})) \left(-s_i I^k[1](s, \mathbf{\Omega}) + I^k[s_i](s, \mathbf{\Omega}) \right). \qquad (9.39)$$

Finally, we obtain as a first-order approximation of the radiative intensity

$$I^k(\mathbf{r}, \mathbf{\Omega}) = I^k(\mathbf{r}, \kappa_k \mathbf{r}, \mathbf{\Omega}) = I_0^k(\mathbf{r}, \kappa_k \mathbf{r}, \mathbf{\Omega}) + \frac{1}{\kappa_k} I_1^k(\mathbf{r}, \kappa_k \mathbf{r}, \mathbf{\Omega})$$

$$= B^k(T_a) + \left(B^k(T(\mathbf{r})) - B^k(T_a) \right) I^k[0](\kappa_k \mathbf{r}, \mathbf{\Omega}) \qquad (9.40)$$

$$+ \frac{1}{\kappa_k} \sum_{i=1}^{3} \frac{\partial B^k}{\partial r_i}(T(\mathbf{r})) \left(-\kappa_k r_i I^k[1](\kappa_k \mathbf{r}, \mathbf{\Omega}) + I^k[s_i](\kappa_k \mathbf{r}, \mathbf{\Omega}) \right).$$

Taking the first moment of (9.40) with respect to $\boldsymbol{\Omega}$, we arrive at the desired expression for the radiative flux vector:

$$\boldsymbol{q}_r(\boldsymbol{r}) = \sum_{k=1}^{M_k} \left(B^k(T(\boldsymbol{r})) - B^k(T_a) \right) \boldsymbol{q}_r^k[1](\kappa_k \boldsymbol{r})$$

$$+ \sum_{k=1}^{M_k} \frac{1}{\kappa_k} \frac{dB^k}{dT}(T(\boldsymbol{r})) \sum_{i=1}^{3} \frac{\partial T}{\partial r_i}(\boldsymbol{r}) \left(-\kappa_k r_i \boldsymbol{q}_r^k[1](\kappa_k \boldsymbol{r}) + \boldsymbol{q}_r^k[s_i](\kappa_k \boldsymbol{r}) \right)$$

where the $\boldsymbol{q}_r^k[\cdot]$ denote the first moments of the corresponding $I^k[\cdot]$ with respect to $\boldsymbol{\Omega}$. The $\boldsymbol{q}_r^k[\cdot]$ are independent of the temperature and therefore have to be calculated only once. This calculation can be carried out for example by ray tracing as presented in section 9.3.2.

Thus, instead of problem (9.1), we have to solve the following problem: $\boldsymbol{r} \in G, \ t \in [0, t^\star]$:

$$c_m \rho_m \frac{\partial T}{\partial t}(\boldsymbol{r}, t) = -\nabla \cdot (-k_h \nabla T(\boldsymbol{r})) +$$

$$- \nabla \cdot \left(\sum_{k=1}^{M_k} \left(B^k(T(\boldsymbol{r})) - B^k(T_a) \right) \boldsymbol{q}_r^k[1](\kappa_k \boldsymbol{r}) \right) +$$

$$- \nabla \cdot \left(\sum_{k=1}^{M_k} \frac{1}{\kappa_k} \frac{dB^k}{dT}(T(\boldsymbol{r})) \sum_{i=1}^{3} \frac{\partial T}{\partial r_i}(\boldsymbol{r}) \left(-\kappa_k r_i \boldsymbol{q}_r^k[1](\kappa_k \boldsymbol{r}) \right. \right.$$

$$\left. \left. + \boldsymbol{q}_r^k[s_i](\kappa_k \boldsymbol{r}) \right) \right)$$

$\boldsymbol{r} \in G$: $T(\boldsymbol{r}, 0) = T_0(\boldsymbol{r})$

with the boundary condition for the heat equation

$$k_h \frac{\partial T}{\partial n}(\boldsymbol{r}_g, t) = \pi \int_{\Lambda_O} [B(T_a(\boldsymbol{r}), \lambda) - B(T(\boldsymbol{r}), \lambda)] d\lambda \qquad (9.41)$$

$$\boldsymbol{r}_g \in \partial G.$$

As can be seen from the numerical simulations presented in the next section, the solution of this new system dramatically improves the Rosseland approximation at virtually no computational cost.

9.4 Numerical Simulation and Results

As a test example, we simulate the cooling of a hot glass cylinder in cold surroundings. We are interested in the temperature distribution inside the glass cylinder during the cooling. Realistic parameters for the glass were given by Schott Glas as described in table 9.1.

The absorption coefficient was approximated by a piecewise constant function as shown in table 9.2.

We have simulated the first 10 seconds. The results are shown in figure 9.3. We compare the temperature distribution calculated with the above discussed methods along the central axis of the cylinder.

The ray tracing of course gives the exact solution. The method requires 30.82 seconds for every time step and needs about 204 MByte memory.

The Rosseland approximation is much faster. Every time step needs only 0.51 seconds. But, as seen from the figure, the simulated temperature profile is not correct. The absorption coefficients for the first wavelength bands are too small. The assumptions upon which the method is based are not valid.

Table 9.1. *Realistic Parameters for Glass*
(from Schott Glas)

Specific heat	c_m	1000	J/kg/K
Density	ρ	2500	kg/m^3
Thermal conductivity	k_h	1	W/m/K
Height of the cylinder	h	0.01	m
Radius of the cylinder	r	0.10	m
Temperature of the surroundings	T_a	20	°C
Initial temperature	T_0	600	°C
Refractive index of glass	n_g	1.46	

Table 9.2. *Absorption Coefficients*

λ_k in μm	λ_{k+1} in μm	κ_k in 1/m
0.01	0.20	0.4
0.20	3.00	0.5
3.00	3.50	7.70
3.50	4.00	15.45
4.00	4.50	27.98
4.50	5.50	267.98
5.50	6.00	567.32
6.00	7.00	7136.06
7.00	∞	opaque

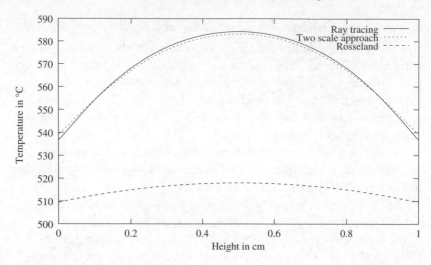

Fig. 9.3. Temperature along the height of the cylinder at time $t = 10$ seconds

Much better results are achieved using our two-scale approach. The temperature profiles of the ray tracing and the two-scale method are nearly identical. Furthermore, the improved method is as fast as the Rosseland approximation (0.55 seconds per time step) and requires only about 4 MByte memory.

9.5 Conclusions and Further Questions

Using a two-scale analysis, a diffusion approximation method has been developed which is efficient and sufficiently accurate for realistic three-dimensional problems. It has been shown that, for a practical example, the diffusion method based on a two-scale analysis is nearly as accurate and 50 times faster than the ray-tracing technique and needs only 1/50 of the memory. The Rosseland approximation which is commonly used in industrial applications can give totally wrong results.

There are a lot of questions which occur relating to extensions of the method and to applications. Two of these are:

- In the present study, we assumed black body boundary conditions for the radiative transfer. How must the described methods be modified for diffuse or specular reflecting boundary conditions?

- We considered the cooling process of glass where scattering was neglected. For some special glasses or ceramics, radiative transfer has to be described not only by absorption and emission but also by scattering. For isotropic scattering, instead of (9.5) we obtain

$$\mathbf{\Omega} \cdot \nabla I^k(\mathbf{r}, \mathbf{\Omega}) + \gamma_k I^k(\mathbf{r}, \mathbf{\Omega})$$

$$= \kappa_k B^k(T(\mathbf{r})) + \frac{\sigma_k}{4\pi} \int\limits_{S^2} I^k(\mathbf{r}, \mathbf{\Omega}') d\mathbf{\Omega}',$$

where σ_k is the scattering coefficient an $\gamma_k = \kappa_k + \sigma_k$.
How can the discussed methods be generalized in this situation?

References

[1] Ames, W. F. (1992) *Numerical Methods for Partial Differential Equations*, Academic Press, Boston, USA.

[2] Choudhary, M. K. & Huff, N. T. (1997) Mathematical modeling in the glass industry: An overview of status and needs. Glastech. Ber. Glass Sci. Technol. **70**, 363–370.

[3] Fiveland, W. A. (1987) Discrete ordinate methods for radiative heat transfer in isotropically and anisotropically scattering media. ASME J. Heat Transfer, **109**, 809–812.

[4] Fiveland, W. A. (1991) The selection of discrete ordinate quadrature sets for anisotropic scattering. Fundamentals of Radiative Heat Transfer, ASME 89–96.

[5] Fiveland, W. A. & Jessee, J. P. (1993) A finite element formulation of the discrete-ordinates method for multidimensional geometries. HTD-Vol.244, Radiative Heat Transfer: Theory and Applications, ASME 41–48.

[6] Howell, J. R. (1988) Thermal radiation in participating media: the past, the present, and some possible futures. Trans. ASME **110**, 1220–1229.

[7] Klar, A. & Siedow, N. (1998) Boundary layers and domain decomposition for radiative heat transfer and diffusion equations: applications to glass manufacturing process. Euro. J. Appl. Math. **9**, 351–372.

[8] Koch, R., Krebs, W., Wittig, S. & Viskanta, R. (1995) Discrete ordinates quadrature schemes for multidimensional radiative transfer. J. Quant. Spectrosc. Radiat. Transfer **53**, 353–372.

[9] Lentes, F.-T. & Siedow, N. (1999) Three-dimensional radiative heat transfer in glass cooling processes. Glastech. Ber. Glass Sci. Technol. **72**, 188–196.

[10] [10] Modest, M. F. (1974) Two-dimensional radiative equilibrium of a gray medium in a plane layer bounded by gray nonisothermal walls. ASME J. Heat Transfer **96C**, 483–488.

[11] Modest, M. F. (1993) *Radiative Heat Transfer*, McGraw-Hill.

[12] Müller, I. (1985) *Thermodynamics*, Pitman, Boston.

[13] Rosseland, S. (1924) Note on the absorption of radiation within a star. M.N.R.A.S. **84**, 525–545.

[14] Siegel, R. & Howell, J. R. (1992) *Thermal Radiation Heat Transfer*, 3rd edition, Taylor & Francis Inc., USA.

[15] Viskanta, R. & Anderson, E.E. (1975) Heat transfer in semitransparent solids. Adv. Heat Transfer **11**, 318–441.

[16] Zingsheim, F. (1999) Numerical solution methods for radiative transfer in semitransparent media. PhD Thesis, University of Kaiserslautern, Germany.

Helmut Neunzert, Norbert Siedow, and Frank Zingsheim

Institute for Industrial Mathematics (ITWM), Erwin-Schrödinger-Straße, D-67663 Kaiserslautern, Germany

neunzert@itwm.uni-kl.de

siedow@itwm.uni-kl.de

zingsheim@itwm.uni-kl.de

10

Water Equilibration in

Vapor-Diffusion Crystal Growth

Preface

A physical and mathematical model is developed for the study of the vapor-diffusion process of protein crystal growth. This is a process widely used to grow high-quality crystals of proteins for the rapidly expanding fields of structural biology and drug design. In this process, an aqueous solution of protein is dynamically concentrated via a passive evaporation from a water drop to a reservoir. The kinetics of the process greatly influence the quality of the crystalline phase, albeit in an unknown manner and with the underlying physical aspects little understood. The model is solved analytically using the method of multiple timescales, identifying and exploiting the disparity in the timescales associated with the various transport mechanisms. Full non-linear transient numerical simulations are also performed and compared with the analytical results and the data obtained from a benchmark experiment. The roles of the controlling parameters in the process are identified, and the requirements for experimental repeatability are explored, especially with regard to the temperature boundary conditions. Finally, it is proposed to use the verified analytical solution for process optimization to reduce the number of independent process variables under consideration.

10.1 Introduction

Structural biology is an emerging-late 20th century science, combining biology, biochemistry, computational science, applied mathematics, and other disciplines, to explicitly decipher the operational characteristics of biomolecules. Hereafter the term protein will be used to describe large conformationally flexible biomolecules that are made by the genetic sequence. Proteins are the building blocks of life, and are responsible for the thousands of individual biochemical actions in living things. Proteins are transcribed from the genetic material using amino acids that are combined via chemical bonds (peptide bonds) to form large molecules. Since each chemical

199

bond could rotate, and since the sequence of amino acids (primary sequence) results in a complex interaction with the water-based solvent surrounding the biomolecule, the overall three-dimensional structure is a unique feature of each protein. The three-dimensional structure is to a large extent responsible for the biological action of a protein. Thus, if it is known, the action of a protein could potentially be understood. Additionally, the treatments of many diseases require that an interaction be established with a part of the protein called the active site in order to produce (or hinder) a specific action. Knowledge of the three-dimensional structure of a specific protein should greatly help in designing small chemical compounds to interact with its active site.

Present methods are insufficient for prediction of the three-dimensional conformation of a protein from its primary sequence. Indeed, this problem is often referred to as the next grand challenge of computational sciences. Current methods of structure determination are experimental in nature. Techniques such as x-ray diffraction are well developed procedures, but they require large single protein crystals with a high degree of internal order. Using these techniques, x-ray diffraction patterns are used to decipher the underlying three-dimensional configuration of the atoms of the protein. Typical requirements are for a protein crystal that is free of structural defects and has a size of a fraction of a millimeter. To date there exists no systematic crystal growth methodology capable of producing consistent high-quality crystals. Growers must instead rely on haphazard screening techniques, in a very large multi-dimensional parameter space, in an attempt to isolate successful preliminary crystallization parameters from the myriad of possibilities.

Of the many techniques used to cultivate crystals [6, 7], the most commonly used on earth and in space is vapor diffusion. The vapor-diffusion process utilizes a drop of protein solution and an aqueous reservoir, both of which are contained in a closed system and influence each other only through an intervening air gap as illustrated in figure 10.1. A precipitating agent such as NaCl is present in both the drop and reservoir, typically in a 1:2 ratio. The agent reduces the vapor pressure over each of the air–liquid interfaces in proportion to its concentration. Thus a water vapor gradient is established in the air gap with a higher water vapor concentration present at the drop–air interface. Water evaporates from the drop, is transported to the reservoir and condenses. During the evaporation process, the involatile solutes in the drop (protein, salt, and buffers) accumulate, thereby reducing the vapor pressure. Evaporation continues until chemical equilibrium is achieved (the vapor pressure over the drop is equal to the vapor pressure over the reservoir). If the initial experiment parameters were properly chosen, the drop protein concentration reaches a supersaturated state allowing crystal nucleation and growth to occur.

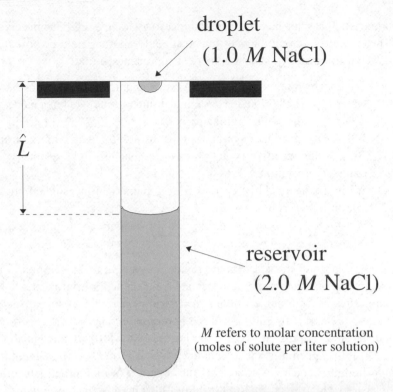

Fig. 10.1. Schematic of the experimental setup

The experimental parameters (drop size and shape, drop–reservoir separation distance, temperature, pH, nature and concentration of all solutes in the drop and reservoir, ambient pressure, and gravitational effects) combine in a complex fashion to dynamically determine the instantaneous vapor pressure gradient between the drop and the reservoir. This gradient dictates the evaporation rate which in turn determines the rate of solute accumulation within the drop (the approach to a supersaturated state). It is important to understand and quantify the evaporation process and its effect on protein concentration, because the approach to supersaturation dictates the process kinetics and thus directly influences the quality of crystals produced. In crystallization experiments conducted by Luft and DeTitta [3], initial and final conditions were held constant while the supersaturation approach was varied. A slower approach was found to yield significantly better crystals with respect to both size and composition.

By quantifying the evaporation process and studying the underlying transport mechanisms, one could potentially determine the key experimental parameters that must be controlled in order to guarantee reproducible crystal growth results. Among the questions that could appropriately be posed and answered using analytical and numerical approaches are:

- What is the minimum set of transport mechanisms for the temperature, water vapor and salt in the system that must be included in a model for accurate understanding of the process?
- What is the relative importance of the different transport mechanisms?
- How do the results vary with respect to changes in the geometry and boundary conditions?
- What lessons could be derived from a numerical investigation of the relevant physics in the process?
- What specific recommendations could be derived to allow experimenters to design better and more consistent processes?

The purpose of this study is to provide answers to these questions via a systematic examination of a representative experimental configuration. The study relies on both numerical and analytical approaches to synergistically examine the problem. Full large-scale numerical models of the experiment were developed from a customized version of Nekton [10], a commercial code based on the spectral-element technique, that is, a method for solving partial differential equations combining the accuracy of high-order polynomial expansions with the geometric flexibility of finite elements. The numerical model included all the relevant physics involved in the evaporation process, including the dynamics of the shrinking water drop. The analytical asymptotic model comes into play in concert with the numerical simulations on its own merits, but more importantly, as a unique tool for subsequent studies. Its principal limitation lies in the set of simplifications that must be set a priori. However, once the asymptotic model is verified, in this case against numerical simulations and also directly against experimental data, it can provide an invaluable tool for process optimization. Whereas a full numerical simulation for a fixed set of parameters may take from 10 to 60 minutes on an advanced workstation (ca. 25 MFLOPS), the equivalent asymptotic model may require no more than one minute of CPU time on a personal computer (ca. 5 MFLOPS). This advantage translates into an ability to perform a large number of parametric studies in the multidimensional parameter space. Moreover, multidimensional optimization typically requires many function evaluations (process simulations). Clearly a fast function evaluation should allow for rapid process optimization. Even advanced design methodologies such as inverse analyses (e.g. determining the proper set of initial conditions to result in a prescribed kinetic path for solution supersaturation) could be approached with a verified asymptotic model at hand.

The experimental benchmark study used for comparison with the results from the analytical and numerical models was conducted by Luft et al. [2]. The experiment focuses on the effect of NaCl (a salt commonly used as a precipitating agent by crystal growers) on the vapor pressure and does not include protein. Drops of 1.0 M (molar concentration, M, measures the number of moles of solute per liter of solution) sodium chloride solution were placed on clear plastic label tape attached to washers. These were sealed over test tubes containing various volumes of 2.0 M sodium chloride solution (figure 10.1). By adjusting the reservoir volume, separation distances ($\hat{L}$) from 0.76 cm to 7.83 cm were achieved. (Typically distances realizable in vapor diffusion crystallizations are on the order of 1 cm.) The system was put in a thermal enclosure, keeping temperature variation to a minimum ($\pm 0.5°C$). At specified time intervals ranging from 20 to 121 hours, the drops were retrieved and their salt content analyzed by refractometry.

The chapter is divided into the following sections. In section 10.2, we present the formulation of the problem, including the governing equations and initial and boundary conditions. Section 10.3 outlines the analytical treatment of the problem, detailing the unique features of the multiple-timescale asymptotic analysis conducted in this work. Section 10.4 summarizes the numerical aspects of the model. We present the results from both models in section 10.5 and discuss the issues raised above in section 10.6.

10.2 Formulation

The models focus on three representative drop–reservoir separation distances used in the experiment [2]: 0.76 cm, 3.81 cm and 7.83 cm, with a coordinate system as shown in figure 10.2. The analysis to follow concentrates on the mass, momentum, and energy transport within the drop and air gap. The reservoir is taken as a region of constant properties. Further, as in the experiments [5], we are examining the approach to equilibrium salt concentration in the drop. The protein phase is not included in the model.

The continuity equation, the Navier–Stokes form of the momentum equation and the energy and mass transport equations are posed axi-symmetrically in the drop and air gap regions. These are

$$\nabla \cdot u = 0, \tag{10.1}$$

$$\rho \left(\frac{\partial u}{\partial t} + u \cdot \nabla u \right) = -\nabla p + \mu \nabla^2 u - \rho(\beta_T(T - T_0) + \Sigma \beta_c c)g, \tag{10.2}$$

Fig. 10.2. Coordinate system used for the model

$$\rho c_p \left(\frac{\partial T}{\partial t} + u \cdot \nabla T \right) = k \nabla^2 T, \tag{10.3}$$

$$\rho \left(\frac{\partial c}{\partial t} + u \cdot \nabla c \right) = \rho D \nabla^2 c, \tag{10.4}$$

where u is the velocity vector, ρ is the density, p is the pressure, μ is the dynamic viscosity, β_T is the volumetric expansion coefficient due to temperature, β_c is the volumetric expansion coefficient due to species c ($\beta_c c$ is summed over all species present), g is the gravity vector, c_p is the specific heat, k is the thermal conductivity, and D is the binary diffusion coefficient of the appropriate species in the appropriate phase. In the drop region, the variables refer to water and c refers to the concentration of salt. In the air gap region, the variables will be given a subscript of a, and c_a refers to the water vapor concentration. Equation (10.3) is solved in the drop phase in the numerical model but not in the air gap region, as will be discussed below.

It should be noted that the above set of governing equations contains for completeness the fluid velocity components. The remainder of the study concentrates on the simplified case corresponding to diffusive transport alone (set $u = 0$ everywhere). Thus the complete set of equations is presented here to illustrate the governing equations necessary for future studies in which the effect of convection will be included.

Due to evaporation, the boundary of the drop is a moving free surface. Rather than pose a complicated fluid mechanics description of this surface together with associated contact conditions between the surface and the label tape, we take a simpler approach. Let V_{int} be the velocity of the interface due to evaporative loss of water. Then the mass flux of water that evaporates from the drop is ρV_{int}. This same amount of water, now in the form of water vapor, diffuses away from the drop at the rate $-\rho_a D_a (\partial c_a / \partial n)$. From a balance between the amount of water that evaporates from the drop and the amount of water vapor that diffuses away from the drop, we define

$$V_{int} = \frac{J}{\rho} = \frac{-\rho_a D_a \frac{\partial c_a}{\partial n}}{\rho}, \tag{10.5}$$

where J is the water vapor flux normal to the interface. Using this definition of the interface velocity, we define the location of the drop interface by posing that the drop radius is given by $r = R - \int_0^t V_{int} dt$. Here R denotes the initial drop radius.

The numerical model assumes the ideal case of an evaporating drop which maintains a hemispherical shape. Picknett and Bexon [11] identify two extreme modes of drop evaporation, fixed contact angle (contact angle remains unchanged while contact area between drop and surface decreases), and fixed base (contact area between drop and surface remains the same while the contact angle diminishes). They conclude that over the drop half-life (the time for one half the initial mass to evaporate), there is no significant difference in the rate of change of drop mass for the two modes of evaporation, particularly when the initial contact angle is large (the initial contact angle is $90°$ in the experiment). We independently verified this behavior with numerical models which pinned the drop end points, allowing it to evaporate in the fixed-contact-area mode. It was also experimentally observed by Luft [2] that slightly oval drops (deformed upon inversion over the reservoir) showed no tendency to equilibrate at different rates than drops which had maintained their hemispherical shapes.

The governing equations are solved subject to the following initial conditions. Initially temperature is constant throughout the entire domain (drop and vapor phases), and the initial salt concentration is uniform within the drop:

$$T = 22°C, \ c = \hat{c}_0 = 0.05844. \tag{10.6}$$

In the vapor phase, we start with the solution for steady-state diffusive transport, $\nabla^2 c_a = 0$, for water vapor based on fixed water vapor values at the drop and reservoir (determined from their respective salt concentrations and the temperature).

Boundary conditions are as follows. At the tape support, $z = 0$, no mass flux and either a fixed temperature or an insulated condition are imposed:

$$\frac{\partial c}{\partial z} = 0, \ \frac{\partial c_a}{\partial z} = 0, \ T = 22°C \ \text{or} \ \frac{\partial T}{\partial z} = 0. \tag{10.7}$$

At the test tube wall, $r = \hat{R}$, no mass flux is imposed:

$$\frac{\partial c_a}{\partial r} = 0. \tag{10.8}$$

Symmetry conditions are imposed at the axis of symmetry, $r = 0$:

$$\frac{\partial T}{\partial r} = 0, \ \frac{\partial c}{\partial r} = 0, \ \frac{\partial c_a}{\partial r} = 0. \tag{10.9}$$

Along the reservoir surface, $z = \hat{L}$, vapor pressure is modeled as a constant, since the amount of water which condenses at the reservoir is minimal in comparison to the reservoir volume and does not significantly alter the solute concentration. (It was verified experimentally that the reservoir NaCl concentration did not measurably change even after multiple equilibrations using a single reservoir.) Thus,

$$c_a = \hat{c}_{res} = 0.01492 \tag{10.10}$$

For a first-order analysis, heat transfer is numerically calculated only in the liquid drop phase with heat transport through the vapor phase approximated by convection from a sphere to an infinite medium, giving an interface heat transfer coefficient $h_c = \text{Nu} k_a / d$, where d is the drop diameter and the Nusselt number (a dimensionless heat transfer coefficient) is equal to its lower limit, $\text{Nu} = 2$, for stagnant air [8]. At the drop interface, conduction in the liquid phase is balanced with convection in the vapor phase (the Newton boundary condition) and the evaporative heat flux

$$-k\frac{\partial T}{\partial n} = h_c(T - T_e) + q_{evap}, \tag{10.11}$$

where T_e is the reference temperature at infinity and the evaporation heat requirement is evaluated from the water flux normal to the interface and the enthalpy of vaporization (H)

$$q_{evap} = -\rho_a D_a \frac{\partial c_a}{\partial n} H. \qquad (10.12)$$

A slightly different formulation of the heat transfer is used in the analytical model and will be explained in section 10.3. Due to the evaporation of water at the interface a salt concentration gradient is established at this front. A mass balance at the drop interface for salt concentration gives

$$-\rho D \frac{\partial c}{\partial n} = \rho c V_{int}. \qquad (10.13)$$

Next we develop a relationship

$$c_a = f(c, T), \qquad (10.14)$$

between the mass fraction of water vapor at the interface, the mass fraction of salt in the drop at the interface, and the temperature of the drop. We note that the vapor pressure in the air gap between the drop and reservoir is sensitive to changes in temperature. Figure 10.3 is taken from data in Mills [8] (pp. 1165). A curve fit to the data is

$$p_0 = 4.02T^2 - 16.76T + 1040.74, \qquad (10.15)$$

where p_0 is the vapor pressure over pure water and T is the temperature. Note that this relation is valid for temperatures ranging from 17 to 24°C, which encompasses the majority of vapor diffusion experiments (generally run at room temperature). We shall see that decreases in temperature due to cooling

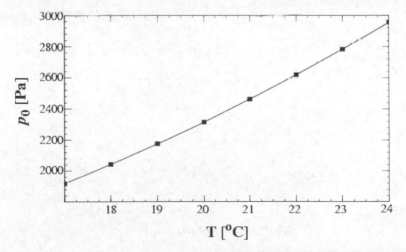

Fig. 10.3. Polynomial approximation to experimental data for vapor pressure over pure water

upon evaporation lead to lower vapor pressures which in turn reduce the rate of evaporation. We now examine the influence of salt on the vapor pressure. Figure 10.4 is based on data from Robinson [12] showing the activity of water (a) for molal NaCl concentrations (m) ranging from 1 to 6 moles per kilogram of solvent. A curve fit to the data is

$$a = -1.4625 \times 10^{-3} m^2 - 3.1275 \times 10^{-2} m + 1. \tag{10.16}$$

In the above, conversion from salt mass fraction to molal concentration is accomplished via the formula

$$m = \frac{1000c}{M_s (1 - c)}, \tag{10.17}$$

where M_s is the molecular weight of salt. The activity is determined by the partial pressure of water vapor over the solution (p_a) and the pressure of water vapor over the pure solvent (p_0) as follows (see Robinson [12] p. 25):

$$a = \frac{p_a}{p_0}. \tag{10.18}$$

Additionally, the mole fraction of water vapor in the air adjacent to the drop interface can be expressed as (see Mills pp. 812–813 [8])

$$x = \frac{p_a}{p_{atm}}, \tag{10.19}$$

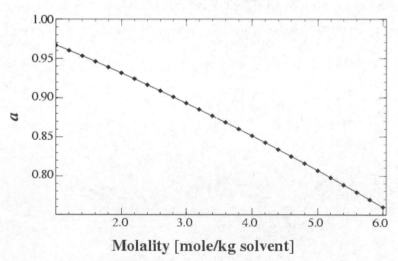

Fig. 10.4. Polynomial approximation to experimental data for water activity of NaCl solutions

where x is the mole fraction of water vapor and p_{atm} is the atmospheric pressure (101,325 Pa). Coupling (10.18) with (10.19) gives

$$x = \frac{ap_0}{p_{atm}}. \tag{10.20}$$

Mole fraction of water vapor is converted to mass fraction for use in the model by the relation [8]

$$c_a = \frac{xM_w}{xM_w + (1-x)M_a}, \tag{10.21}$$

where M_w and M_a are the molecular weights of water and air, respectively. Substituting equations (10.15), (10.16), and (10.20) into equation (10.21) leads to the relationship posed in equation (10.14). Hence, higher salt concentrations lead to smaller activity values which imply lower amounts of water vapor. The result is that the loss of water in the drop due to evaporation concentrates the salt in the drop and thus reduces the vapor pressure difference between the drop and the reservoir.

10.3 Analytical Treatment

10.3.1 Geometry

There is one further simplification we make in the model formulation. From experimental observation, Luft [2] concluded that the equilibration rates of drops which had deformed on inversion did not deviate significantly from those of hemispherical drops. We take this as an indication that drop shape may have negligible influence on the equilibration rates. Hence, in order to define a separable domain for the mathematical solution, we replace the hemispherical drop pool by a cylindrical pool of equal volume, as depicted in figure 10.5. In this formulation, the radius of the drop is fixed to be the radius of the test tube, and the free surface is flat. The location of the free surface will be given by

$$z = h - \int_0^t V_{int} dt. \tag{10.22}$$

Here z is the axial coordinate and h denotes the initial drop height.

An alternative approach to achieving a separable domain is to replace the cylindrical test tube by a "spherical" tube as shown in figure 10.6. This is the geometry used by Fowlis [1]. Unfortunately in this geometry the reservoir has an infinite surface area at infinity. Thus, there is always a water vapor gradient,

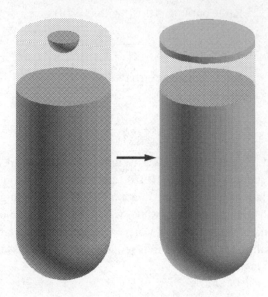

Fig. 10.5. Schematic of actual to cylindrical setup

and evaporation will take place even if the reservoir is infinitely far from the drop. Hence, equilibration will take place in a finite time. This is contrary to one's intuition in the cylindrical test tube case, where one expects that, if the reservoir is infinitely far away, then equilibration will require infinite time. This may partially explain why the Fowlis [1] results do not agree with the Luft [2] experimental data.

10.3.2 Method of Multiple Timescales

In the development of the model to follow, our concern is to simulate the rate at which the water equilibration is achieved. Hence our drop solution is composed of water and 1.0 M NaCl. No protein macromolecule is present in the solution. Further, we neglect convection within the drop. While it is true that evaporation will leave behind a boundary layer rich in the heavier salt, in the cylindrical formulation this does not lead to an unstable density configuration, because the heavier salt layer sits at the bottom of the drop along the drop interface. This is in contrast to a spherical drop. There the heavier salt layer along the drop interface will sink along the interface towards the drop bottom due to buoyancy.

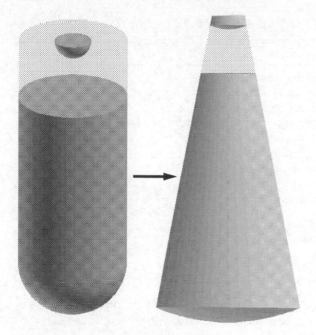

Fig. 10.6. Schematic of actual to spherical setup

Under these assumptions, we identify three primary timescales within the hanging drop system. The fastest timescale is that associated with the diffusion of water vapor across the gap between the drop and reservoir. This scale is

$$t_1 = \frac{\hat{L}^2}{D_a}, \tag{10.23}$$

where $\hat{L}$ is the distance between the reservoir and label tape; t_1 is on the order of 10 s using the data listed in table 10.1. The intermediate timescale is that associated with the diffusion of salt within the drop. This scale is

$$t_2 = \frac{R^2}{D}, \tag{10.24}$$

which is on the order of 10^4 s (i.e. hours). The slowest timescale is that associated with the change in vapor pressure at the drop interface. From (10.5), we can identify a characteristic value, V, for V_{int}:

$$V = \frac{D_a \rho_a c^*}{\rho R}. \tag{10.25}$$

Table 10.1. *Reference Data for the Hanging Drop System*

Parameter	Value
Radius of the test tube $\hat{R}$	6.8×10^{-3} m
Radius of the hemispherical drop R	2.2×10^{-3} m
Height of the cylindrical drop h	1.535×10^{-4} m
Distance from reservoir to label tape $\hat{L}$	0.0076–0.0783 m
Diffusivity of salt in water D	1.48×10^{-9} m^2/s
Diffusivity of water vapor in air D_a	2.5×10^{-5} m^2/s
Density of water ρ	1000 kg/m^3
Density of air ρ_a	1.1985 kg/m^3
Viscosity of water μ	9.68×10^{-4} kg/m/s
Viscosity of air μ_a	1.8225×10^{-5} kg/m/s
Thermal conductivity of water k	0.602 W/m/K
Thermal conductivity of air k_a	0.0264 W/m/K
Specific heat of water c_p	4181 J/kg/K
Specific heat of air c_{pa}	1006 J/kg/K
Enthalpy of vaporization H	2.449×10^6 J/kg
Molecular weight of water M_w	18 kg/kmole
Molecular weight of air M_a	29 kg/kmole
Molecular weight of salt (NaCl) M_s	58.44 kg/kmole
Mass fraction $\hat{c}_{Res}$ of water vapor in air at reservoir	0.01492
Initial salt concentration in drop $\hat{c}_0$	1 mol/liter
Initial salt concentration in drop $\hat{c}_0$	0.05844 (mass fraction)
Initial drop temperature T_d	295.15 K
Reference temperature at infinity T_e	295.15 K

Here c^* represents the initial change in mass fraction of water vapor between the drop and reservoir. Using (10.14), c^* is calculated as

$$c^* = f(\hat{c}_0, T_d) - \hat{c}_{Res}, \qquad (10.26)$$

where $\hat{c}_0$ is the initial salt mass fraction in the drop (equation (10.6)(b)), T_d is the initial drop temperature, and $\hat{c}_{Res}$ is the mass fraction of water vapor in the air at the reservoir (equation (10.10)). Using V, we define a timescale for the evaporation,

$$t_3 = \frac{R}{V}, \qquad (10.27)$$

on the order of 10^5 s, which is tens of hours. We now define the parameter ϵ as the time ratio

$$\epsilon = \frac{t_2}{t_3} = \frac{VR}{D}, \quad \tau = \epsilon t. \qquad (10.28)$$

Thus ϵ is on the order of 10^{-1} and will serve as the small parameter in the asymptotic analysis to follow. Also t_2 will be the timescale. We introduce the slow timescale, $\tau = \epsilon t$, and use the method of multiple scales [9].

In the multiple timescale procedure, we shall assume that t_2 is the relevant $O(1)$ timescale. On this timescale, we shall see that processes taking place on the t_1 timescale have equilibrated. In order to capture the cumulative behavior of processes occurring on the t_3 timescale, the multiple timescale procedure introduces τ as an additional independent variable.

10.3.3 Formulation

We nondimensionalize the system by scaling all lengths with h and time with t_2. All temperatures are scaled with T_d, and all mass fractions are scaled with c^*. These scalings lead to several nondimensional groups. The magnitudes of these groups will be related to the size of ϵ, since ϵ is the small parameter in the analysis. We have

$$\frac{D}{D_a} = \hat{D}\epsilon^4, \tag{10.29}$$

where $\hat{D}$ is an $O(1)$ quantity. In essence the diffusivity ratio is on the order of 10^{-4}, as given by the data in table 10.1, so that diffusion of water vapor in air is much faster than diffusion of salt in water. Additional groups are

$$\frac{D}{\kappa_a} = Le_a\epsilon^2, \tag{10.30}$$

$$\frac{D}{\kappa} = Le. \tag{10.31}$$

Here κ_a and κ are the thermal diffusivities for air and water, respectively. Le_a and Le are Lewis numbers and are taken as $O(1)$ quantities. The final two groups characterize the cooling due to evaporation. We have

$$\frac{H}{T_d c_p} = E, \tag{10.32}$$

$$\frac{k_a}{k} = K\epsilon. \tag{10.33}$$

E and K are $O(1)$ quantities. E measures the energy lost by the drop due to evaporation, and the rate of this loss is proportional to V_{int} via equations (10.5) and (10.12). Since V_{int} is $O(\epsilon)$, then the rate of this loss is $O(\epsilon)$. On the other hand there is a warming effect due to the presence of the reservoir. This heat transfer is also $O(\epsilon)$ as defined by the conductivity ratio $K\epsilon$. Hence the orderings listed in (10.32) and (10.33) lead to a balance between the various energy transfer effects in the system.

In the multiple timescale procedure, the chain rule is used to replace the $\partial/\partial t$ derivatives in the original governing equations with $(\partial/\partial t) + \epsilon(\partial/\partial \tau)$. Under these multiple timescale and scaling assumptions, and neglecting convection, the governing equations for diffusive salt and energy transport within the cylindrical drop are

$$\frac{\partial c}{\partial t} + \epsilon \frac{\partial c}{\partial \tau} = \frac{\partial^2 c}{\partial r^2} + \frac{1}{r}\frac{\partial c}{\partial r} + \frac{\partial^2 c}{\partial z^2}, \tag{10.34}$$

$$\text{Le}\left(\frac{\partial T}{\partial t} + \epsilon \frac{\partial T}{\partial \tau}\right) = \frac{\partial^2 T}{\partial r^2} + \frac{1}{r}\frac{\partial T}{\partial r} + \frac{\partial^2 T}{\partial z^2}. \tag{10.35}$$

These equations are valid in the drop domain, $O \leq r \leq \tilde{R} = \hat{R}/h, 0 \leq z \leq 1 + \epsilon \int_0^t (\partial c_a/\partial z)dt$. We note that we have used (10.5) and (10.22) to define the drop interface, $z = 1 + \epsilon \int_0^t (\partial c_a/\partial z)dt$. Under the small ϵ assumption, evaporation is slow, so that the interface location also changes slowly.

The governing equations for water vapor and energy transport in the air gap between the drop and reservoir are

$$\epsilon^4 \hat{D}\left(\frac{\partial c_a}{\partial t} + \epsilon \frac{\partial c_a}{\partial \tau}\right) = \frac{\partial^2 c_a}{\partial r^2} + \frac{1}{r}\frac{\partial c_a}{\partial r} + \frac{\partial^2 c_a}{\partial z^2}, \tag{10.36}$$

$$\epsilon^2 Le_a\left(\frac{\partial T_a}{\partial t} + \epsilon \frac{\partial T_a}{\partial \tau}\right) = \frac{\partial^2 T_a}{\partial r^2} + \frac{1}{r}\frac{\partial T_a}{\partial r} + \frac{\partial^2 T_a}{\partial z^2}. \tag{10.37}$$

These equations are valid in the air gap domain, $O \leq r \leq \tilde{R}, 1 + \epsilon \int_0^t (\partial c_a/\partial z)dt \leq z \leq L = \hat{L}/h$. The reservoir is assumed to fill the domain, $O \leq r \leq \tilde{R}, L \leq z < \infty$. The reservoir is assumed to have a fixed temperature T_d, and a fixed salt mass fraction so that $c_{Res} = \hat{c}_{Res}/c^*$ remains constant. Hence the reservoir enters the mathematical model only through boundary conditions at the reservoir surface, $z = L$.

The boundary conditions for the above equations are the following. Along the test tube sidewall, $r = \tilde{R}$, we have no mass flux and no heat transfer:

$$\frac{\partial c}{\partial r} = \frac{\partial c_a}{\partial r} = \frac{\partial T}{\partial r} = \frac{\partial T_a}{\partial r} = 0. \tag{10.38}$$

At the label tape, $z = 0$, we pose no mass flux and either a fixed temperature or an insulated boundary condition

$$\frac{\partial c}{\partial z} = 0, \ T = 1 \text{ or } \frac{\partial T}{\partial z} = 0. \tag{10.39}$$

The latter conditions will allow us to bound the magnitude of heat transfer between the drop and label tape. At the reservoir, $z = L$, we have

$$c_a = c_{Res}, \quad T_a = 1, \tag{10.40}$$

as discussed earlier. Finally, at the drop interface, $z = 1 + \epsilon \int_0^t (\partial c_a / \partial z) dt$, we have the following mass conditions which result from (10.13) and (10.14):

$$\frac{\partial c}{\partial z} + \epsilon \frac{\partial c_a}{\partial z} c = 0, \tag{10.41}$$

$$c_a = \frac{f(c^* c, T_d T)}{c^*}. \tag{10.42}$$

Further we have the energy conditions

$$T = T_a, \tag{10.43}$$

$$\frac{\partial T}{\partial z} = \epsilon Le E \frac{\partial c_a}{\partial z} + \epsilon K \frac{\partial T_a}{\partial z}. \tag{10.44}$$

The two terms on the right-hand side of the latter equation represent cooling due to evaporation and warming due to heat transfer across the air gap. As discussed earlier, both processes are $O(\epsilon)$.

Finally, we have the initial conditions at $t = 0$:

$$c = \tilde{c}_0 = \frac{\hat{c}_0}{c^*}, \quad T = 1. \tag{10.45}$$

10.3.4 Solution

For the governing equation (10.34)–(10.45), we seek solutions for the dependent variables in the form

$$F(r, z, t, \tau) = F_0(r, z, t, \tau) + \epsilon F_1(r, z, t, \tau) + \cdots, \tag{10.46}$$

where F represents c, c_a, T, and T_a. Substituting these forms into the governing equations, we have the leading-order problem

$$\frac{\partial c_0}{\partial t} = \frac{\partial^2 c_0}{\partial r^2} + \frac{1}{r} \frac{\partial c_0}{\partial r} + \frac{\partial^2 c_0}{\partial z^2}, \tag{10.47}$$

$$Le \frac{\partial T_0}{\partial t} = \frac{\partial^2 T_0}{\partial r^2} + \frac{1}{r} \frac{\partial T_0}{\partial r} + \frac{\partial^2 T_0}{\partial z^2}, \tag{10.48}$$

$$0 = \frac{\partial^2 c_{a0}}{\partial r^2} + \frac{1}{r} \frac{\partial c_{a0}}{\partial r} + \frac{\partial^2 c_{a0}}{\partial z^2}, \tag{10.49}$$

$$0 = \frac{\partial^2 T_{a0}}{\partial r^2} + \frac{1}{r} \frac{\partial T_{a0}}{\partial r} + \frac{\partial^2 T_{a0}}{\partial z^2}. \tag{10.50}$$

The boundary conditions for the above equations are the following. Along $r = \tilde{R}$, we have no mass flux and no heat transfer:

$$\frac{\partial c_0}{\partial r} = \frac{\partial c_{a0}}{\partial r} = \frac{\partial T_0}{\partial r} = \frac{\partial T_{a0}}{\partial r} = 0. \tag{10.51}$$

At the label tape, $z = 0$, we have

$$\frac{\partial c_0}{\partial z} = 0 \tag{10.52}$$

and

$$T_0 = 1 \quad \text{or} \quad \frac{\partial T_0}{\partial z} = 0. \tag{10.53}$$

At the reservoir, $z = L$, we have

$$c_{a0} = c_{Res}, \quad T_{a0} = 1. \tag{10.54}$$

At the drop interface, $z = 1$, we have the following:

$$\frac{\partial c_0}{\partial z} = 0, \quad c_{a0} = \frac{f(c^* c_0, T_d T_0)}{c^*}, \quad T_0 = T_{a0}, \quad \frac{\partial T_0}{\partial z} = 0. \tag{10.55}$$

Finally, we have the initial conditions at $t = 0$:

$$c_0 = \tilde{c}_0, \quad T_0 = 1. \tag{10.56}$$

The solutions to the above equations are

$$c_0 = C(\tau), \tag{10.57}$$

$$T_0 = \theta(\tau), \tag{10.58}$$

$$c_{a0} = \frac{f(c^* C(\tau), T_d \theta(\tau))}{c^*} + \frac{c_{Res} - (f(c^* C(\tau), T_d \theta(\tau)))/c^*}{L - 1}(z - 1), \tag{10.59}$$

$$T_{a0} = \theta(\tau) + \frac{1 - \theta(\tau)}{L - 1}(z - 1). \tag{10.60}$$

In the above equations, $C(\tau)$ and $\theta(\tau)$ are unknown functions of τ which will be determined at the next order. From equations (10.56)(a) and (10.56)(b) we have $C(0) = \tilde{c}_0$ and $\theta(0) = 1$. If we had not introduced the multiple scale technique, the leading-order solutions would have been these constant initial conditions. Hence, in order to reach equilibrium (the drop salt concentration changing from 1 mole/liter to 2 mole/liter), the $O(\epsilon)$ corrections would need to be of the same size as the leading-order solutions. This violates the concept of uniformly valid solutions. Hence there is a need to introduce the multiple timescales because, even though evaporation is slow, it occurs over a long period of time, and thus has an important cumulative effect.

Examining the leading-order solutions, one must keep in mind that these correspond to small ϵ, which means slow evaporation. Hence the drop is characterized by the spatially uniform salt and temperature distributions $C(\tau)$ and $\theta(\tau)$. In other words, the diffusion of salt and temperature to spatially uniform distributions is rapid compared to the evaporation. Further, if we return to equation (10.5) and the nondimensionalization scheme and use (10.59), we have that

$$V_{int} = -V\frac{\partial c_{a0}}{\partial z} = V\hat{V}(\tau), \tag{10.61}$$

where

$$\hat{V}(\tau) = -\frac{c_{Res} - (f(c^*C(\tau), T_d\theta(\tau)))/c^*}{L - 1}. \tag{10.62}$$

From these expressions we note that, as $L \to \infty$, then $V_{int} \to 0$ as one expects for the reservoir at ∞. We also see that $V_{int} \to 0$ as $f(c^*C(\tau), T_d\theta(\tau))/c^* \to c_{Res}$, which is the condition for equilibrium. Finally, we comment that evaporation is taking place since

$$c_{Res} \lessgtr \frac{f(c^*C(\tau), T_d\theta(\tau))}{c^*}.$$

Now, to determine $C(\tau)$ and $\theta(\tau)$, we must go to $O(\epsilon)$. We account for the $O(\epsilon)$ terms in the boundary conditions (10.41) and (10.44) by introducing the transformations

$$c_1(r, z, t, \tau) = C_1(r, z, t, \tau) + \frac{1}{2}z^2C(\tau)\hat{V}(\tau), \tag{10.63}$$

$$T_1(r, z, t, \tau) = \theta_1(r, z, t, \tau) + \frac{1}{2}z^2\left(LeE\frac{\partial c_{a0}}{\partial z} + K\frac{\partial T_{a0}}{\partial z}\right). \tag{10.64}$$

Using these transformations, we find that, at $O(\epsilon)$, the boundary conditions (10.41) and (10.44) become

$$\frac{\partial C_1}{\partial z} = \frac{\partial \theta_1}{\partial z} = 0, \tag{10.65}$$

while the governing equations for salt and energy transport at $O(\epsilon)$ are

$$\frac{\partial C_1}{\partial t} + \frac{\partial C}{\partial \tau} - C(\tau)\hat{V}(\tau) = \frac{\partial^2 C_1}{\partial r^2} + \frac{1}{r}\frac{\partial C_1}{\partial r} + \frac{\partial^2 C_1}{\partial z^2}, \tag{10.66}$$

$$Le\left(\frac{\partial\theta_1}{\partial t}+\frac{\partial\theta}{\partial\tau}\right)-\left(LeE\frac{\partial c_{a0}}{\partial z}+K\frac{\partial T_{a0}}{\partial z}\right)=\frac{\partial^2\theta_1}{\partial r^2}+\frac{1}{r}\frac{\partial\theta_1}{\partial r}+\frac{\partial^2\theta_1}{\partial z^2}.$$

$$(10.67)$$

In the above equations, the leading-order solutions appear as inhomogeneous terms within the partial differential equations. These terms are not functions of the independent variables t, r, and z. It can be shown that these inhomogeneous terms give rise to solutions which are multiples of t. Such solutions lead to nonuniformities in the expansions (10.46) for large values of t. To remove these nonuniformities the following amplitude equations must be true in order to achieve solutions which are uniformly valid in time.

$$\frac{\partial C}{\partial\tau}-C(\tau)\hat{V}(\tau)=0,\qquad(10.68)$$

$$Le\frac{\partial\theta}{\partial\tau}-\left(LeE\frac{\partial c_{a0}}{\partial z}+K\frac{\partial T_{a0}}{\partial z}\right)=0,\qquad(10.69)$$

subject to

$$C(0)=\tilde{c}_0,\qquad(10.70)$$

$$\theta(0)=1.\qquad(10.71)$$

This coupled set of equations determines the unknowns $C(\tau)$ and $\theta(\tau)$. These equations are solved numerically using a fourth-order Runge–Kutta scheme. For comparison with experimental results, it is more convenient to rescale the above equations so that time is measured in hours and salt concentration is measured in mole/liter. Making this change, and using (10.15)–(10.21), we find the following coupled set of amplitude equations, which will be solved numerically:

$$\frac{\partial C}{\partial\tau}=\Gamma C\left(\frac{g_3(C)g_1(\theta)}{g_2(\theta)}-0.01502\right),\qquad(10.72)$$

$$\frac{\partial\theta}{\partial\tau}=\Gamma\left(-\frac{H}{c_pT_d}\left(\frac{g_3(C)g_1(\theta)}{g_2(\theta)}-0.01502\right)+\frac{k_a}{D\rho c_p}(1-\theta)\right),\quad(10.73)$$

where

$$\Gamma=\frac{D_a\rho_a3600}{\rho h^2(L-1)},$$

$$g_1(\theta)=1040.74-368.762\theta+1947.52\theta^2,$$

$$g_2(\theta)=28.887-0.211\theta^2+0.0400\theta,$$

$$g_3(C)=0.000178(1-0.0332C-0.00165C^2),$$

subject to

$$C(0) = 1, \tag{10.74}$$

$$\theta(0) = 1. \tag{10.75}$$

10.4 Numerical Approach

The governing equations, together with the initial and boundary conditions, were solved using a spectral-element method. A customized version of Nekton [10] that allowed for interactions among multiple domains was constructed. In the numerical model the shape of the drop was hemispherical at all times and the volume was dynamically determined during the transient nonlinear calculations. Typically 20 macro higher-order elements were used to discretize the domain, with 5×5 collocation points in each. The macro elements preserve the geometrical flexibility of the finite-element method, while the polynomial expansions provide for exponential convergence rates for smooth solutions. The use of variational projection operators and Gauss numerical quadrature results in a set of discrete field equations. These equations are solved using tensor-product sum-factorization methods. The solution algorithms typically rely on iterative preconditioned conjugate-gradient techniques.

Second-order implicit time differencing is used to advance the solution in time. Convergence of the solution at each time step is obtained based on preset criteria for forcing function and solution residuals. The simulations were conducted on a Silicon Graphics R8000 multiprocessor machine with anywhere between 1 and 4 processors dedicated to the solution that typically took from 10 to 60 minutes of wall clock time. Finally, the solution could be graphically rendered at any time step using an advanced graphical user interface running under X-Windows.

10.5 Results

The comprehensive numerical and the simplified asymptotic models represent tools designed to study the vapor-diffusion process with two different aims. The use of a fully coupled numerical model is primarily of value for a detailed examination of the importance of the various transport mechanisms. Further, the numerical results from the more comprehensive model check the assumptions made in deriving the asymptotic model. The analytical asymptotic model was used principally as a fast simulation tool whose utility would be further explored in future optimization studies of actual processes.

We begin by providing in figure 10.7 a direct comparison between experimental data and numerical simulations for three drop–reservoir gaps. These simulations assumed a constant temperature of 22°C at the top of the tape support. As the evaporation process proceeds, the transport of heat from the air and from the base tape support through the drop to the interface counterbalances the cooling due to release of heat of evaporation. The innocent-looking stretched exponential evaporation curve in time is actually the result of a complex interplay of various transport mechanisms. Indeed, while the data for the shortest gap (0.76 cm) show an exponential-like behavior, no such universal behavior is evident for longer gaps. It is also clear that the predicted evaporation rates are faster than the measured rates for the smaller gap size.

Temperature is one parameter in the problem that is shown to have a singular importance. Temperature enters the problem by affecting the vapor pressure (see figure 10.3) on the interface. As discussed above, the interface temperature drops due to the release of heat of evaporation. However, when one considers the smallness of the expected magnitude of the temperature depression (a fraction of a degree) and its concomitant influence on the vapor pressure, its effects are not readily recognizable. It is only after integrating the role of small changes in the vapor pressure over the long time in the experiment that the importance of the temperature effects becomes clear. The overprediction of the evaporation rate for the shortest gap shown in figure 10.7 is most likely due to the incorrect assumption of constant temperature over the experiment. In this case, the faster evaporation rate resulted in cooling of

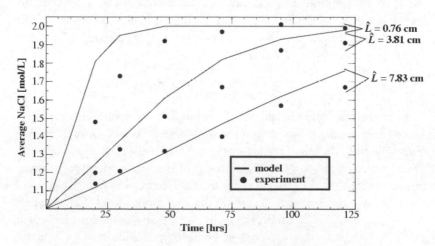

Fig. 10.7. Numerical model vs. experiment for three separation distances

the drop and the support, perhaps below the assumed level of 22°C. As will be demonstrated below, even a small fraction of a degree could result in large variations in the course of the process. The better match with the larger gap data could be explained by the apparent lack of variation of the tape support temperature from the ambient temperature of 22°C commensurate with the slower evaporation rate.

Figure 10.8 demonstrates the small temperature differences that are typical in the experiment. The analytical prediction of a uniform temperature field in space, equation (10.58), is consistent with this result. The minimum temperature on the interface against time is depicted in figure 10.9 for the previously discussed case of a constant tape support temperature and for the extreme case of a perfectly insulated support. In the latter case the interface temperature drops further, since the heat supplied to the interface must be conducted through the air. The large effect of the temperature variation on the interface is demonstrated in figure 10.10, where the two extreme cases (constant tape temperature and insulated tape) are compared with regard to the temporal water flux through the interface. It is clear that an accurate quantification of the temperature boundary condition is a very important parameter in specifying the problem.

Temperature boundary conditions should be more important at smaller gaps, since the rate of change in vapor pressure controlling the evaporation process is strongly dominated by the rate at which heat could be supplied to the interface. Figure 10.11 depicts the two extreme cases of temperature boundary conditions considered (insulated and constant temperature at 22°C) for the smallest and largest gap sizes. The experimental data for the large gap (slowest evaporation rate) is adequately matched by the constant temperature boundary condition. The data for the faster evaporation case lies in between the two extreme cases. It is believed that the faster evaporation and heat flux requirements in that case resulted in a subsequent depression of the tape temperature, since no active control was provided.

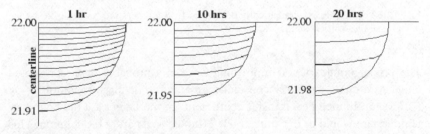

Fig. 10.8. Isotherms for the drop ($\hat{L}$ = 0.76cm) with a boundary condition of 22°C on the tape support. Isotherms are spaced 0.0067°C apart

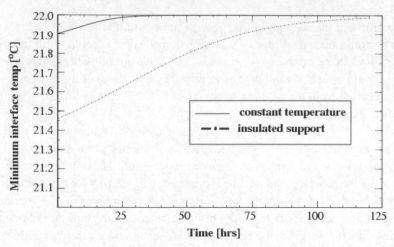

Fig. 10.9. Minimum temperature at the drop interface ($\hat{L}$ = 0.76cm) with constant temperature and insulated boundary conditions on the tape support

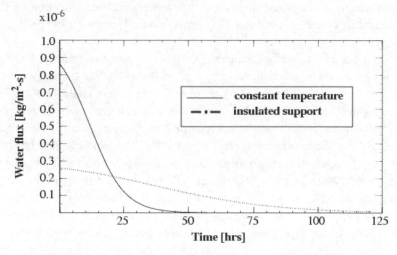

Fig. 10.10. Water flux is dramatically decreased by a temperature change of less than $0.5^{o}C$ (see figure 10.9) on the interface

The proper boundary condition for this case is probably best represented either as a quasi-stationary constant-temperature boundary condition or by a diffusion-limited heat transfer coefficient. In the latter case the boundary temperature would be dynamically determined as a part of the solution. This case could not be addressed in this study, due to lack of available information on the experimental setup beyond the tape support.

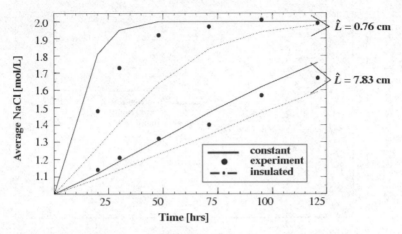

Fig. 10.11. Results from the numerical model with constant temperature and insulated boundary conditions on the tape support

The analytical model discussed in section 10.3 makes several important simplifications and, through a multiple-timescale asymptotic analysis, provides a coupled set of ordinary differential equations in time. Recall that the analytical model differs from the numerical model in two aspects. First of all, the drop is cylindrical rather than hemispherical, and secondly the temperature field within the air gap is calculated rather than approximated by an interfacial heat transfer coefficient. Results from the analytical model ((10.73)–(10.75)) solved in Mathematica [4] are presented in figure 10.12 in comparison with the experimental data. The case designated as isothermal in figure 10.12 corresponds to a fixed temperature field, $\theta(\tau) = 1$ in equation (10.58) throughout the domain. This case should result in the fastest evaporation rates, since the heat of evaporation on the interface could instantaneously be replenished. The second case corresponds to the insulated-tape-support boundary condition, equation (10.39)(c). This case thus allows for heat to be replenished to the interface through diffusion in the air gap from the reservoir below. This case resulted in an overall good agreement with experimental data for the short gap examined. A further comparison of the experimental and analytical results in figure 10.13 shows an overall reasonable agreement. The systematic underprediction at longer gap sizes is apparently due to the conversion of partly hemispherical to cylindrical drop geometry in the analysis, which results in a lower water vapor gradient at the interface, and partly due to the insulated temperature boundary condition on the tape support. The analytical model thus examines the fastest evaporation limit and the case designated to depict the recommended process to be discussed in the following section.

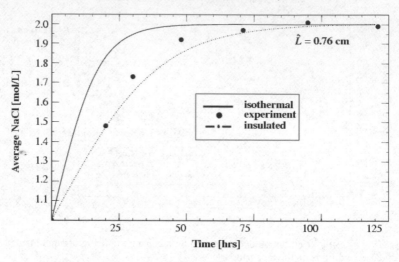

Fig. 10.12. Comparison of analytical and experimental results

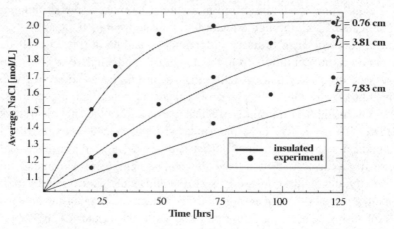

Fig. 10.13. Analytical results for the insulated case compared to experiment for three separation distances

10.6 Discussion

Spontaneous evaporation of water drops is a well-studied topic in the literature, first examined by Maxwell [5]. This study concentrates on a fundamentally different physical process, one that is governed by the nonlinear effects of precipitating agents on the driving force for evaporation.

Analytical models by Fowlis [1] and Sibille [13] lay the groundwork for the vapor-diffusion process. Both separate the evaporation process into five steps: transport of water molecules from the interior of the drop to the interface, evaporation at the drop surface, transport of water vapor to the reservoir, condensation at the reservoir surface and transport of water molecules from the surface of the reservoir to the interior. Evaporation and condensation are assumed to be instantaneous, and characteristic times are developed for transport of water molecules within the drop, within the vapor gap and within the reservoir. They determine the rate-limiting step of the process to be the transport of water through the vapor gap, with a characteristic time $\hat{L}^2/D_a = O(10s)$ for a distance $\hat{L} = 1.2$ cm (a typical separation distance between the drop and the reservoir in a vapor diffusion cell).

Further examination of the relevant timescales shows that transport of water vapor is not the limiting step. Characteristic times for the diffusion of salt within the drop, the diffusion of heat within the drop and the change in vapor pressure at the drop interface can be constructed as follows.
Diffusion of salt:

$$\frac{R^2}{D} = O(10^4 s). \tag{10.76}$$

Diffusion of heat:

$$\frac{R^2}{\kappa} = O(10^2 s), \tag{10.77}$$

Change in vapor pressure:

$$\frac{R}{V_{Int}} = O(10^6 s). \tag{10.78}$$

The drop radius R begins at 2.25×10^{-1} cm, and the interface velocity V_{int} is defined in equation (10.5). The last time constant associated with the change in vapor pressure was not considered in previous analyses. Its inclusion in this analysis appears from an examination of the nondimensional form of the boundary conditions at the drop interface. When the entire process is controlled by the difference in vapor pressure between the drop interface and the reservoir, the time constant at which this difference is changing should be of importance as well. Since the vapor pressure at the reservoir is constant, we consider only the time constant associated with the vapor pressure on the drop. Since the vapor pressure varies with the salt concentration, which in turn increases proportionally to the decrease in the drop volume, it is the time constant associated with the rate of decrease in the drop size that must be evaluated. While the exact value of the interface velocity V_{int} is a priori unknown, it could be estimated by considering the average velocity over the entire duration of the process. For example, for the shortest and longest separation distances (0.76 cm and 7.83 cm, respectively), the average interface velocities were 1.06×10^{-7} cm/s and 8.8×10^{-8} cm/s.

As computed in equation (10.78), it becomes apparent that it is indeed this time constant that controls the dynamics for the majority of cases. For very short separation distances, in which the diffusion of salt from the interface into the drop interior cannot keep up with the rapid rate of evaporation on the interface, a salt boundary layer may form near the interface. This boundary layer would result in a commensurate decrease in the vapor pressure, thus slowing down the process. In this case, consideration of time constants alone does not lead to proper conclusions, and full numerical simulations are necessary to evaluate the contribution of each physical component of the process. In comparison, the timescales for diffusion of salt (for longer separation distances) and heat within the drop are small, but temperature cannot a priori be discarded. The process is controlled by the interface vapor pressure which is a nonlinear function of both the temperature and NaCl concentration.

A revisit of the list of pertinent questions in section 10.1 shows that the results indeed provide information relevant to delineating the important physical aspects of this process. Specifically, it was demonstrated that the entire nonlinear transient set of transport processes must be included for a faithful simulation of this seemingly simple evaporation process. This said, it was also shown that certain simplifications, if judiciously made, provide for a much simplified analytical approach to be constructed. Temperature was determined to be a crucial parameter in the process. Small temperature variations (of the order of a fraction of a degree) were sufficient to change the results significantly. We also were able to determine that the rate-limiting step in the evaporation process is the rate of change of the vapor pressure at the drop interface. The interface vapor pressure is dependent on the salt and temperature at the interface, which are in turn dependent on the rate of change in the drop volume.

In the practice of drop evaporation experiments, one typically varies a certain set of parameters while keeping others fixed. The gap length is typically held constant, as determined by the commonly used Linbro plate on which the experiments are conducted. The drop and reservoir volumes are also typically constant. The process temperature is maintained reasonably constant by placing the apparatus in either a thermostat or in an insulated box. More careful control of the temperature is difficult and not usually done. The major variables in the process are the type of precipitating agents added to the drop, and the rate of evaporation as determined by the initial vapor pressure difference corresponding to the type and relative concentrations of the salts. Since the experimental matrix is constructed to primarily compare the effects of different precipitating agents, it would be highly advisable to remove other parameters of importance in the process kinetics on which one has little direct control. Based on the results presented, it is recommended that a good insulator be placed on top of the tape support to provide for a nearly insulated temperature boundary condition. This simple approach to control would be superior to difficult alternatives such as active thermal control. Heat would then be supplied through the air gap which is typically of constant dimension. The process should allow a systematic comparison among the different agents in the experimental matrix, without the uncontrolled variation due to small differences in the tape support temperature during the experiment.

The use of the simplified analytical model is envisioned for future work in tandem with experiments for optimizing the process. Instead of tracking the individual influence of the many combinations of control variables in the process (such as geometry, gap size, temperature, etc.), the model provides a concise statement on the rate of evaporation dependency on the nondimensional combination of such parameters. The all-important but ill-defined aspect of crystallization in the path kinetics could thus be quantitatively evaluated for each process. A direct application of these models for practical vapor-diffusion process optimization is the goal of future studies.

The study presented here demonstrates the utility of mathematical and numerical modeling to decipher complex physical phenomena. Additional studies are suggested in the following areas:

- Reduction of the analytical results to a single representative non-dimensional group that could be used in design of experiments screening methodologies to represent the entire kinetic evaporation path.
- Adaptation of the models to specific geometrical configurations popular among experimentalists, such as sitting drop, hanging drop, etc.
- Examination of the role of convective transport on the process.
- Inclusion of protein physics in the drop; in particular, prediction of supersaturation kinetics and probability of crystal nucleation.

Acknowledgement

This work was supported by NASA Grant NCC-3-494.

References

[1] Fowlis, W., DeLucas, L., Twigg, P., Howard, S., Meehan, E. & Baird, J. (1988) Experimental and theoretical analysis of the rate of solvent equilibration in the hanging drop method of protein crystal growth. J. Crystal Growth **90**, 117–129.

[2]   Luft, J., Albright, D., Baird, J. & DeTitta, G. (1996) The rate of water equilibration in vapor-diffusion crystallizations: Dependence on the distance from the droplet to the reservoir. Acta Cryst. D **52**, 1098–1106.

[3]   Luft, J. & DeTitta, G. (1997) Kinetic aspects of macromolecular crystallization. Methods in Enzymology **276**, 110–131.

[4]   *Mathematica* (1997) Wolfram Research, Inc., Champaign, IL, USA.

[5]   Maxwell, J. C. (1877) Diffusion, *Encyclopedia Brittanica*.

[6]   McPherson, A. (1982) *Preparation and Analysis of Protein Crystals*, John Wiley & Sons.

[7]   McPherson, A. (1993) Virus and protein crystal growth on earth and in microgravity. J. Phys. D **6**, B104–112.

[8]   Mills, A. F. (1995) *Heat and Mass Transfer*, Irwin Publishing, Toronto, Canada, 288–289.

[9]   Nayfeh, A. H. (1981) *Introduction to Perturbation Techniques*, Wiley, NY.

[10] *Nekton*, (1997) Nektonics Inc., Cambridge, MA, USA.

[11] Picknett, R. & Bexon, R. (1977) The evaporation of sessile or pendant drops in still air. J. Colloid and Interface Sci. **61**, 336–350.

[12] Robinson, R. & Stokes, R. (1955) *Electrolyte Solutions*, Academic Press Inc., NY, USA, 461–462.

[13] Sibille, L. & Baird, J. (1991) Solvent evaporation rates in the closed capillary vapor diffusion method of protein crystal growth. J. Crystal Growth **110**, 80–88.

Arnon Chait, Elizabeth Gray

Computational Microgravity Laboratory, NASA Glenn Research Center, Cleveland, Ohio 44135, USA

Arnon.Chait@grc.nasa.gov

lizabeth@cml-mail.grc.nasa.gov

Gerald W. Young

Department of Mathematics and Computer Science, The University of Akron, Akron, Ohio 44325-4002, USA

gwyoung@uakron.edu

11

Modelling of Quasi-Static and Dynamic

Load Responses of Filled

Viscoelastic Materials

11.1 Introduction

The engineering uses of rubber have expanded well beyond traditional products such as tires and seals. Today elastomeric, or rubber-like, components can be found in a diverse set of constructs including engine mounts, building foundations, belts, and fenders (see [13], [22]). Increasingly, the applications of rubber are becoming more sophisticated, as exemplified by the use of rubber bearings in bridges which allow for thermal expansions of the deck without placing excessive loads on the bridge supports (see [15]).

In current engineering applications, elastomer composites are typically filled with inactive particles such as carbon black or silica. If active fillers were used, such as piezoelectric, magnetic, or conductive particles, the result-ing controllable elastomer could be used in products such as active or smart vibration suppression devices (e.g. see [8], [17], [18]). As these new materials are developed, the role of design will increase in both complexity and importance. In particular, the capability to predict the dynamic mechanical response of the components will become increasingly valuable.

The many desirable characteristics of rubber as a design component, which include the ability to undergo large elastic deformations and provide significant damping with near incompressibility, are also contributing factors to the complications arising in the process of formulating models. Damping is highly complex, and the strain history, rate of loading, environmental temperature, and amount and type of filler affect the mechanical response in a nontrivial manner. Additionally, many elastomers exhibit strong hysteresis characteristics similar to those found in shape-memory alloys and piezoceramic actuators. Hysteretic effects in an elastomer include a time-history-dependent stress–strain constitutive law.

Lord Corporation, a company based in Cary, NC, develops and manufactures a wide variety of damping devices, including many types of elastomer-based dampers. In 1994, applied mathematicians with the Center for Research in Scientific Computation (CRSC) at North Carolina

State University began collaborations with scientists at Lord Corporation to develop a high-resolution model of the dynamic behavior of elastomers. The ultimate goal of this project is to construct a dynamic model and accompanying design tools that will assist Lord in the development of higher quality damping devices. Specifically, scientists at Lord were interested in gaining a better understanding of the dynamic response of elastomeric materials to external loads and deformations. The original team members included H.T. Banks, N.J. Lybeck, Y. Zhang and N. Medhin from the CRSC, and M.J. Gaitens, B.C. Muñoz, and L.C. Yanyo from Lord Corporation.

Our team has concentrated on two types of deformation: simple extension and simple shear. Since most complex motion is believed to be a combination of these two types of motion (see [10], [23]), a basic understanding of each is an important precursor to a full understanding of complex deformations. Moreover, simple shear and simple extension investigations are more easily pursued than those for most other types of motion in that experimental data is readily obtained. Lord Corporation is well equipped to collect data for these two types of motion, and these data are currently used by engineers in product design. Here we will give a summary of our efforts to date on simple extension.

The first two years of the team's collaboration included an extensive study of existing elastomer models. In spite of the many complex issues relating to elastomer dynamics, researchers have made substantial progress in developing tools for such models (see [14], [23], [27], [28] for basic texts). These models are predominantly phenomenological, based on strain energy function (SEF) and finite-strain (FS) theories. SEF theories contain information about the elastic properties of elastomers, but do not describe either damping or hysteresis, and hence are typically used for *static*, or time-independent finite-element analysis (see [10]). The CRSC/Lord team worked, both theoretically and computationally, with several of these models. Computations based on the existing models led to the development of more general nonlinear models. Dynamic and quasi-static experiments were designed and carried out to test the performance of the models on different types of elastomer. Details on these models and experiments will be outlined in section 11.2.

In working with the existing and generalized nonlinear models, the team recognized the need for a model that could incorporate the effects of hysteresis and damping. Additional experiments were conducted, and the quasi-static case was considered for developing a nonlinear, hysteretic stress–strain constitutive law. G.A. Pintér and L.K. Potter of the CRSC joined the team in 1996 as hysteresis was being incorporated into both the quasi-static and dynamic models. The formulation of the dynamic model led to the question of well-posedness, which was addressed and resolved. Finally, additional computations were performed to validate the accuracy of the model on several types of elastomers. These results will be presented in section 11.3.

11.2 Nonlinear Extension Models, Experiments and Results

11.2.1 Neo-Hookean Extension Models

The strain energy function (SEF) and finite-strain (FS) models can be used as a first step in the development of a dynamic model for the behavior of elastomers. These models are designed for materials that are usually assumed to be isotropic and incompressible; the simplest of these are known as neo-Hookean materials.

The SEF material models use the principal extension ratios λ_i to represent the deformed length of unit vectors parallel to the principal axes (the axes of zero shear stress). The SEF models of Mooney and Rivlin are based on Rivlin's proposal [24] that the SEF should depend only on the strain invariants $I_1 = \lambda_1^2 + \lambda_2^2 + \lambda_3^2$, $I_2 = \lambda_2^2\lambda_3^2 + \lambda_1^2\lambda_3^2 + \lambda_1^2\lambda_2^2$ and $I_3 = \lambda_1^2\lambda_2^2\lambda_3^2$. For example, the Mooney SEF is given by $U = C_1(I_1 - 3) + C_2(I_2 - 3)$, or more generally, the modified expression $U = C_1(I_1 - 3) + f(I_2 - 3)$, where C_1 and C_2 are constants, and f has certain qualitative properties. This expression is most appropriate for components where the rubber is not tightly confined and where the assumption of absolute incompressibility (implying $\lambda_1\lambda_2\lambda_3 = 1$ or $I_3 = 1$) is a reasonable approximation. The more general Rivlin SEF $U = \sum_{i+j>1}^{N} C_{ij}(I_1 - 3)^i(I_2 - 3)^j$ and its generalization for near incompressibility (see [10]) allow higher-order dependence of the SEF on the invariants. The works of Ogden [23], as well as Valanis and Landel [10], represent an important departure from Rivlin's proposal, in that their formulations are based on SEFs that depend only on the extension ratios.

The finite-strain elastic theory of Rivlin [24], [28] is developed with a generalized Hooke's law in an analogy to infinitesimal-strain elasticity, but requires no "small deformation" assumption and includes higher-order exact terms in its formulation. Moreover, finite stresses are defined relative to the *deformed body* and hence are the "true stresses", as opposed to the "nominal" or "engineering" stresses (relative to the *undeformed body*) one usually encounters in the infinitesimal-strain linear elasticity used with metals. This Eulerian measure of strain is an important feature of any development of models for use in combined analytical, computational and experimental investigations of rubber-like material bodies. The finite-strain elasticity of Rivlin can be directly related to the strain energy function formulations through equations relating the finite strains $\tilde{e}_{xx}$, $\tilde{e}_{yy}$, $\tilde{e}_{zz}$ to the extension ratios λ_1, λ_2, λ_3 used to define the SEF.

The finite-strain approach can be put in a somewhat more general perspective in the context of classical modelling of elastic solids and fluids (filled elastomers do not fit exactly into either category, although in the absence of cross links, elastomers are highly viscous fluids). In classical approaches, one frequently encounters an Eulerian formulation in dynamics of fluids, where large deformations or displacements are common, whereas a Lagrangian formulation is employed for solids undergoing small elastic deformations. In both formulations momentum balance laws along with constitutive laws relating stress and strain are employed to develop theories of dynamics (see [23], [21]). In the general Lagrangian formulations, quantities (such as stress, strain) are defined relative to an original or reference configuration $\mathcal{B}_0$ of the body or structure in terms of a fixed coordinate system $X = \{X_i\}$. For an Eulerian formulation, one defines quantities relative to a "current" or deformed configuration $\mathcal{B}$ with coordinates $x = \{x_i\}$ relative to the deformed configuration.

A fundamental role in discussing the relationship between these formulations is a "configuration" map $x = \chi(X)$ or "motion" $x(t) = \chi(t, X)$ if the deformations are changing in time from an original configuration $\chi(0, X) = X$. The deformations (in the usual elasticity terminology) are then given by

$$u(t, X) = \chi(t, X) - \chi(0, X) = \chi(t, X) - X.$$

The configuration gradient (also called the "deformation gradient" in an unfortunate misnomer) is defined by

$$A = \frac{\partial \chi(X)}{\partial X} = \frac{\partial x}{\partial X}$$

and is used to define the right ($A^{\mathrm{T}}A$) and left (AA^{T}) Cauchy–Green "deformation tensors" and the usual (in elasticity theory) Green–St. Venant strain

$$\mathcal{E} = \frac{1}{2}(A^{\mathrm{T}}A - I) = \frac{1}{2}(D^{\mathrm{T}}D + D + D^{\mathrm{T}}),$$

where $D = \partial u/\partial X = A - I$. These definitions, along with momentum balance laws, can be used to write dynamic equations in either the Lagrangian (reference) or Eulerian (current) coordinate systems. If one uses the Eulerian system, stresses are given in terms of the Cauchy or true stress $\mathbf{T}$ and constitutive (stress–strain) laws are expressed in terms of current coordinates x. For computational and experimental purposes, however, it is often more desirable to use a Lagrangian coordinate system, and then the Cauchy or true stress must be converted to an expression for the nominal or engineering stress $S = JA^{-1}T$, where $J = \det A$ and S^T is called the First Piola–Kirchhoff stress tensor.

The SEF or finite-strain formulations can be used along with standard material independent force and moment balance derivations (the Timoshenko theory [26], [12]) as the basis of a dynamic model for extension. Here we take a simple example: an isotropic incompressible ($\lambda_1\lambda_2\lambda_3 = 1$) elastomeric rod with a tip mass under *simple elongation* with a finite applied stress in the principal axis direction $x_1 = x$ (see figure 11.1). Following standard convention, we use lowercase letters to denote the Lagrangian coordinates. The finite-stress theory (or the Mooney theory with SEF $U = C_1(I_1 - 3)$) leads for *neo-Hookean* materials to a true stress $T = \frac{E}{3}(\lambda_1^2 - (1/\lambda_1))$, or an engineering or nominal stress

$$\sigma = \frac{T}{\lambda_1} = \frac{E}{3}\left(\lambda_1 - \frac{1}{\lambda_1^2}\right),$$

where, in terms of deformation u in the x direction, we have, since deformations in the y and z directions are negligible in simple extension,

$$\lambda_1^2 = 1 + 2\tilde{e}_{xx}$$

$$= 1 + 2\frac{\partial u}{\partial x} + \left(\frac{\partial u}{\partial x}\right)^2 = \left(1 + \frac{\partial u}{\partial x}\right)^2,$$

or $\lambda_1 = 1 + (\partial u/\partial x)$. Note in this case that the stress tensors T and S reduce to one nontrivial component $T = T_{11}$ and $\sigma = S_{11}$. Here E is a generalized modulus of elasticity, and we note that these formulations are restricted to $\lambda_1 > 1$.

This can be used in the Timoshenko theory for longitudinal vibrations of a rubber bar with a tip mass, to obtain

$$\rho A_c \frac{\partial^2 u}{\partial t^2} + \gamma \frac{\partial u}{\partial t} - \frac{\partial \sigma}{\partial x} = 0 \quad 0 < x < \ell,$$

$$M \frac{\partial^2 u}{\partial t^2}(t, \ell) = -\sigma \Big|_{x=\ell} + F(t) + Mg,$$

(11.1)

where ρ is the mass density, $F(t)$ is the applied external force, A_c is the cross-sectional area, M is the tip mass, g is the gravitational constant, γ is

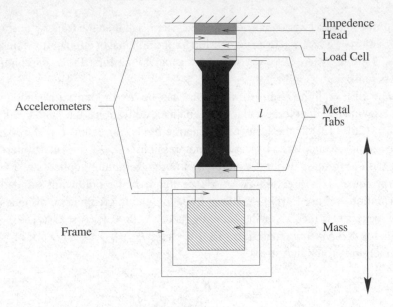

Fig. 11.1. Rod with tip mass under tension

the air damping coefficient, and

$$\sigma = \frac{A_c E}{3}\left(\lambda_1 - \frac{1}{\lambda_1^2}\right) + A_c C_D \frac{\partial \lambda_1}{\partial t}$$

$$= \frac{A_c E}{3}\tilde{g}\left(\frac{\partial u}{\partial x}\right) + A_c C_D \frac{\partial^2 u}{\partial t \partial x} \tag{11.2}$$

is the internal (engineering) stress resultant, with $\tilde{g}(\xi) = 1 + \xi - (1 + \xi)^{-2}$. Here we have included a Kelvin–Voigt damping term (C_D is the Kelvin–Voigt damping coefficient) in the stress σ as a first attempt to model damping. This leads to the nonlinear partial differential equation initial-boundary-value problem (defined for $0 < x < \ell$)

$$\rho A_c \frac{\partial^2 u}{\partial t^2} + \gamma \frac{\partial u}{\partial t} - \frac{\partial}{\partial x}\left(\frac{E A_c}{3}\tilde{g}\left(\frac{\partial u}{\partial x}\right) + A_c C_D \frac{\partial^2 u}{\partial t \partial x}\right) = 0, \tag{11.3}$$

$$M\frac{\partial^2 u}{\partial t^2}(t, \ell) = -\left(\frac{A_c E}{3}\tilde{g}\left(dudx\right) + A_c C_D \frac{\partial^2 u}{\partial t \partial x}\right)\Bigg|_{x=\ell} + F(t) + Mg,$$

$$u(t, 0) = 0,$$

$$u(0, x) = \varphi_0,$$

$$u_t(0, x) = 0,$$

for dynamic longitudinal displacements of a neo-Hookean material rod in extension. Since a series expansion of $\tilde{g}$ for small ξ yields $\tilde{g}(\xi) = 3\xi - 3\xi^2 + 4\xi^3 - \cdots$, this is readily seen, in the case of small displacements, to reduce to the usual longitudinal deformation system for Hookean materials.

In general, one does not expect the initial-boundary-value problem associated with (11.3) to have a classical (smooth) solution. Equations such as (11.3) involve infinite-dimensional state space models [8] (i.e., the state space is a vector space which is *not* finite-dimensional), and must be discretized before the solution can be approximated computationally. We have chosen to use a Galerkin method with linear splines for the spatial discretization. The second-order equation (11.3) is then written as a first-order system in time, and a stiff equation solver is used for the time integration.

As reported in [3], the form of $\tilde{g}$ for (11.3) obtained by either the neo-Hookean or simple Mooney SEF is inadequate in capturing the behavior of most filled elastomers. An *essential* task then is to determine a more general nonlinearity $\tilde{g}$ with the aid of experiments and inverse problem techniques.

11.2.2 Approximation of Nonlinear Constitutive Laws

The neo-Hookean model (11.3) provides a natural example of a nonlinear PDE in the modelling of elastomers, but has only limited practical application since it is inadequate in describing most filled elastomers. Typically, one would employ equations such as (11.2) with a more general nonlinearity $\tilde{g}$, which should be estimated from experiments. One does not expect such a general $\tilde{g}$ to admit a SEF as a function of either the strain invariants or the extension ratios. Comparisons with SEF methods can be made by using the (approximate) SEF to derive the expected stress–strain relationship, and comparing results in the stress–strain plane. With this goal in mind, we now proceed to discuss the numerical estimation of $\tilde{g}$.

11.2.2.1 Quasi-Static Experiments and Inverse Problems

Although our ultimate goal is to use the results of dynamic experiments to determine the nonlinearity $\tilde{g}$, a reasonable first step is to estimate $\tilde{g}$ using data from quasi-static pull tests. Quasi-static motion includes time-dependent behavior, but we require that there be zero acceleration. While this type of experiment cannot be used to study damping, it can be used to study the nonlinearity $\tilde{g}$, and more generally the constitutive stress–strain law.

With this in mind, we designed and implemented quasi-static pull tests on an Instron machine at Lord Corporation. A cylindrical rubber sample is

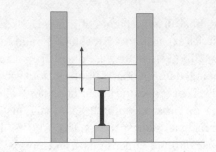

Fig. 11.2. Quasi-static pull test

secured vertically into the test machine, so that the lower end of the rod
($x = 0$) is fixed, and the upper end of the rod ($x = \ell$) is attached to a
load cell and a horizontal crossbar (see figure 11.2). A displacement pattern
$\Delta(t)$ for the upper end of the rod is programmed into the machine, and the
resulting loading force $f(t)$ is recorded. In these experiments, we initially
used a constant rate of displacement at 5 inches per minute.

Using the quasi-static load and displacement data at $x = \ell$, an inverse problem can be formulated to estimate the nonlinearity $\tilde{g}$. In general, this type of
inverse problem is infinite-dimensional both in state and parameter space, and
hence finite-dimensional approximations must be made. Preliminary results
suggested a nonlinear but piecewise linear form (e.g. *linear splines*) might
perform well in approximating the function $\tilde{g}$. This type of approximation for
$\tilde{g}$ was able to capture the load and displacement curves for unfilled natural
rubber, reconfirming the need for a nonlinear $\tilde{g}$ (see [3] for details).

After obtaining satisfactory results for the quasi-static case, we may use
the resulting form for $\tilde{g}$ as an initial estimate in algorithms for determining
$\tilde{g}$ with dynamic data. Dynamic motion is more general than the quasi-static
motion that we described above, in the sense that we allow for all orders of
time derivatives. We next describe these results.

11.2.2.2 Dynamic Experiments and Inverse Problems

With the aid of scientists at Lord Corporation, we designed and carried out
dynamic free-release and impulse hammer experiments with various samples
of elastomers. These two types of experiments were chosen not only for the
relative simplicity of the model for simple extension; it is well known that
even in simple engineering materials (e.g. a spring–mass–dashpot system), it
is easier to study dissipation of energy from free release or impulse response
data than from other types of experiments such as cyclical tests (the almost
universal dynamic tests currently reported in rubber research literature).

In these experiments, a cylindrical elastomeric rod is suspended vertically so that the top end ($x = 0$) is fixed, and a tip mass is attached at the lower end ($x = \ell$) as in figure 11.1. The frame at the bottom of the structure was used as an additional mass and as a housing for an accelerometer. An additional accelerometer is placed above the top of the elastomer to verify the clamped boundary condition at $x = 0$. A load cell is attached between the elastomer and the top accelerometer to provide force measurements at $x = 0$.

For the free-release experiment, the frame and rod were supported so that the rod itself was at its natural length (i.e. no compression or extension). The support was then removed, allowing the mass to fall freely. In the impulse response experiment, the rod and mass hung freely until equilibrium was reached, and then a hammer was used to excite the system.

In both experiments described above, we used an unfilled natural rubber, a natural rubber very lightly filled with carbon black, and highly filled natural rubber samples. In this section we summarize results pertaining to the unfilled sample. Using insight gained from the quasi-static inverse problem, we used the following inverse problem formulation to obtain the nonlinearity $\tilde{g}$:

$$\text{Minimize} \quad J(\tilde{g}) = \frac{1}{2} \sum_{i=1}^{M} \left| z_i - A_c \sigma \left(\frac{\partial u}{\partial x}(t_i, 0; \tilde{g}) \right) \right|^2$$

over g in some admissible class of functions $\mathcal{G}$. The observations z_i, obtained from the experiments, are the force measurements at the fixed end, and u is the solution of (11.3) corresponding to $\tilde{g}$.

Two forms for $\tilde{g}$ were chosen for use in the inverse problem. First $\tilde{g}$ was chosen as the best possible linear function, and this was followed by choosing the best $\tilde{g}$ from a class $\mathcal{G}$ of four-term piecewise linear splines (see [3] for details). Figures 11.3 and 11.4 demonstrate that the best linear $\tilde{g}$ does not adequately capture the dynamics of the rubber, while a nonlinear function can give satisfactory results.

These same experiments and inverse problems were repeated for moderately and highly filled rubber, but with less encouraging results. As the amount of filler is increased, the hysteretic behavior of the elastomer is more prevalent and must be included in the model. Since the highly filled elastomers are more frequently used in engineering applications than unfilled rubbers, because of their increased strength and stiffness, an accurate model for the dynamic behavior of highly filled elastomers is most desirable. This led to the need for developing a nonlinear hysteretic constitutive law for filled elastomers.

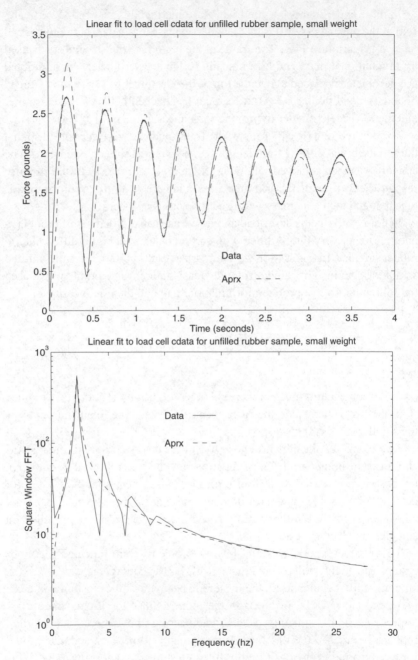

Fig. 11.3. (top) Time domain approximation with a linear $\tilde{g}$, and (bottom) the FFT of the solution and the data

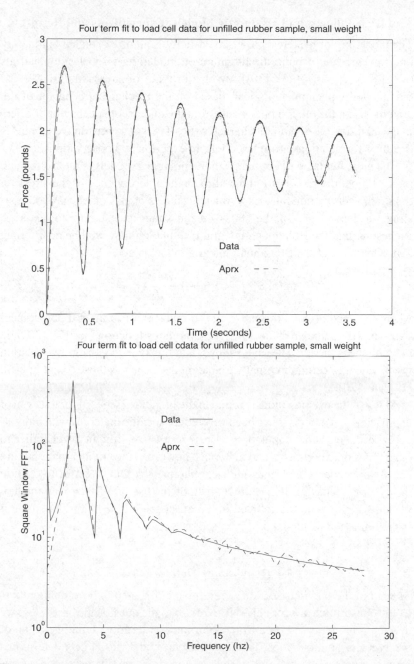

Fig. 11.4. (top) Time domain approximation with a four-term piecewise linear $\tilde{g}$, and (bottom) the FFT of the solution and the data

11.3 Nonlinear and Hysteretic Models, Experiments and Results

The modelling of hysteresis in viscoelasticity is an often studied but largely unresolved issue. Among the literature on modelling viscoelastic materials (e.g. [9], [11], [14], [25], [30]), two major types of stress–strain law exist. One model is phenomenological, based on the mechanical behavior of the materials, and the other model stems from the microscopic behavior of fillers in the material. One common characteristic of existing constitutive models is that their formulations arose largely from experimental observations.

The most fundamental of the phenomenological models is the Boltzmann integral model, which is formulated to capture dependence of the stress on the strain and/or strain rate. Experimental studies of elastomers have demonstrated a dependence of the stress on the strain and strain-rate histories, suggesting that the Boltzmann integral formulation may well be appropriate here. One form of the Boltzmann integral model is given by

$$\sigma(t) = \int_{-\infty}^{t} Y(t-s) \frac{d\epsilon(s)}{ds} ds,$$

where $\sigma(t)$ and $\epsilon(t)$ are the stress and (infinitesimal) strain at time t respectively, and Y is known as the relaxation modulus function (see [7], [30] and the references therein). It can be shown that this type of constitutive relation is equivalent to certain internal variable models, such as those found in [16], [19], and [20].

Many of the existing models found in the literature (see [30] for a survey of approaches and models) have been verified for specific materials with individual experiments and computations. These results are difficult to generalize to other types of experiments and materials, however, since rubber materials are heavily dependent on various physical parameters. Our attempts to capture the dynamic behavior of elastomers include the use of a Boltzmann integral model, which we first developed in the quasi-static case before including it in the full dynamic model.

11.3.1 Quasi-Static Hysteresis Loops

As seen in the previous section, a reasonable first step in developing a hysteretic constitutive law for (11.1) is to return to the quasi-static case. Here we develop a model based on a Boltzmann integral term, with the inclusion of two types of nonlinearities. The form for our model was largely determined by experimental observations of quasi-static stress–strain curves for various samples of elastomers, and our inability to obtain useful fits to data with linear hysteresis models.

One well-known characteristic of rubber-like materials is that they exhibit nested stress–strain curves, known as hysteresis loops (see figure 11.5). This type of data was collected on the Instron machine (as in section 11.2.2.1), and includes a sequence of loading and unloading the sample with progressively smaller maximum strain levels (i.e. the outermost loop is created first, then the inner loops follow in sequence).

Our main approach to modelling is a pseudophenomenological one. We assume that the elastomer's quasi-static behavior is prescribed by both an elastic response and a viscoelastic response. We choose a law of the following form:

$$\sigma(t) = g_e(\epsilon(t)) + \int_{-\infty}^{t} Y(t-s) \frac{d}{ds} g_v(\epsilon(s), \dot{\epsilon}(s)) \, ds, \qquad (11.4)$$

where g_e and g_v are the elastic and viscoelastic response functions respectively, and Y is known as the memory kernel. Experiments and calculations in the quasi-static case led to the conclusion that elastomers have a finite memory, so that the stress depends on the strain and strain rate only for a history of length r. That is, the memory kernel Y obeys $Y(t) \approx 0$ for all $t \geq r$, which

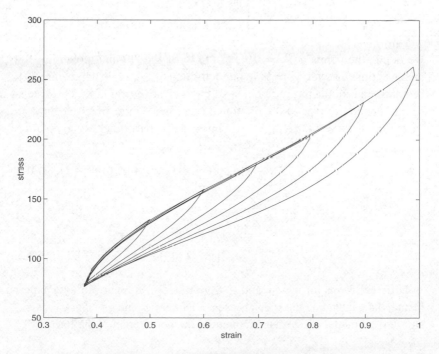

Fig. 11.5. Nested hysteresis loops for rubber highly filled with carbon black

implies that we may approximate (11.4) by

$$\sigma(t) = g_e\left(\epsilon(t)\right) + \int_{t-r}^{t} Y(t-s)\frac{d}{ds}g_v\left(\epsilon(s), \dot{\epsilon}(s)\right) ds. \tag{11.5}$$

For experimental and computational purposes, we further assume that, in each experiment, the motion in the material prior to time $t = 0$ is negligible, and that the duration of the experiment is less than r. Hence we have $(d/dt)g_v\left(\epsilon(t), \dot{\epsilon}(t)\right) \approx 0$ for all $t < 0$. Moreover, (11.5) can be approximated by

$$\sigma(t) = g_e\left(\epsilon(t)\right) + \int_{0}^{t} Y(t-s)\frac{d}{ds}g_v\left(\epsilon(s), \dot{\epsilon}(s)\right) ds. \tag{11.6}$$

The quasi-static stress–strain behavior in figure 11.5 suggests that the elastomer possesses a different viscoelastic response for loading ($\dot{\epsilon} > 0$) than for unloading ($\dot{\epsilon} < 0$). As a result, we choose a piecewise continuous form for the function g_v:

$$g_v\left(\epsilon(s), \dot{\epsilon}(s)\right) = \begin{cases} g_{vi}\left(\epsilon(s)\right) & \text{for } \dot{\epsilon}(s) > 0, \\ g_{vd}\left(\epsilon(s)\right) & \text{for } \dot{\epsilon}(s) < 0. \end{cases}$$

We define the points t_k $(k = 1, \ldots, M)$ to be the "turning points," or the points in time for which $\dot{\epsilon} = 0$. In our definition of g_v, we do not require g_v to be continuous at the turning points, so the derivatives in (11.4)–(11.6) must be interpreted in a type of distributional sense made more precise below.

Assuming $t_K < t < t_{K+1}$, we can integrate by parts in (11.6) to obtain

$$\sigma(t) = g_e\left(\epsilon(t)\right) + \int_{0}^{t} \dot{Y}(t-s)g_v\left(\epsilon(s), \dot{\epsilon}(s)\right) ds + Y(0)g_v\left(\epsilon(t), \dot{\epsilon}(t)\right)$$

$$- Y(t)g_{vi}\left(\epsilon(0)\right) + \sum_{k=1}^{K} Y(t-t_k)(-1)^{k+1}$$

$$\times \left[g_{vi}\left(\epsilon(t_k)\right) - g_{vd}\left(\epsilon(t_k)\right)\right]. \tag{11.7}$$

Different forms for g_e, g_v, and Y were tested in comparison with experimental data, including the special cases of a linear g_e and g_v, with $g_{vi} \equiv g_{vd}$. Our current model utilizes an exponential memory kernel of the form

$$Y(t) = C_2 e^{-C_1 t},$$

and we have

$$g_e(x) = \sum_{i=1}^{3} E_i x^i,$$

$$g_{vi}(x) = \sum_{i=1}^{3} a_i x^i, \qquad g_{vd}(x) = \sum_{i=1}^{3} b_i x^i,$$

where $C_1, C_2 > 0$ and E_i, a_i, b_i $(i = 1, 2, 3)$ are real constants. A detailed discussion on the development of these functions can be found in [30].

The constants C_1, C_2 and E_k, a_k, b_k are material-dependent parameters that must be determined using parameter estimation techniques. This is implemented by formulating an inverse problem based on quasi-static experimental data. Using the Instron as before, we obtain displacement and force observations $(\Delta(t_i), f(t_i))$, $(i = 1, \ldots, N)$ at the point $x = \ell$. These observations can be converted into infinitesimal strain $\hat{\epsilon}(t_i)$ and stress $\hat{\sigma}(t_i)$ according to $\hat{\epsilon} = \Delta/\ell$ and $\hat{\sigma} = f/A_c$, where A_c is the original cross-sectional area of the sample.

We estimate the parameters C_1, C_2, E_k, a_k, b_k, $(k = 1, 2, 3)$, which are collectively denoted as q, by fitting this stress–strain data in the following inverse problem:

$$\text{Minimize} \qquad J(q) = \frac{1}{2} \sum_{i=1}^{N} \left| \hat{\sigma}(t_i) - \sigma(t_i; q) \right|^2 \qquad (11.8)$$

over q in some admissible parameter set Q, where $\sigma(t_i; q)$ denotes the quantity obtained by inserting the data $\hat{\epsilon}(t_i)$ into (11.7) with the parameters q.

We carried out the computations in MATLAB and FORTRAN, using both BFGS-type and Nelder–Mead optimization routines. This inverse problem was solved using data from several types of elastomer, including silica-filled silicone and lightly, moderately and highly filled carbon black. Here we present results for the highly filled carbon black sample (the type of elastomer most important to industry).

Our main goal in solving an inverse problem is to find parameters q^* that will best predict a set of nested hysteresis loops like those found in figure 11.5. If possible, we would like to use data from at most one or two of the loops in the inverse problem itself, yet still obtain an accurate prediction of the entire set. After experimenting with different data sets, we chose the data from the 100% and 90% strain loops (the two outer-most loops in figure 11.5) for use in the inverse problem. That is, we let $\{\hat{\epsilon}(t_i), \hat{\sigma}(t_i)\}$, $i = 1, \ldots, N$ in (11.8) be the strain and stress data from the 100% and 90% strain loops, and we solved

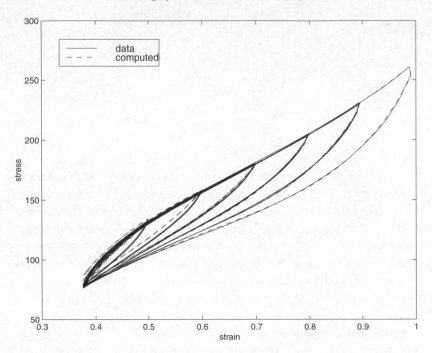

Fig. 11.6. Nested hysteresis loops for rubber highly filled with carbon black: data and model predicted curves

the inverse problem for (11.8) to obtain an optimal parameter set q^*. We then used these parameters in (11.7) to simulate stress–strain curves and compare with data for the inner loops. The resulting curves are shown in figure 11.6, with a relative error of 2.2% between model prediction and data.

Equivalent results for the other elastomer samples were similarly encouraging, leading us to conclude that the constitutive relation (11.7) provides satisfactory predictions of quasi-static stress–strain data for several types of elastomer. Since our ultimate goal is to accurately predict the *dynamic* behavior of elastomer, we return to the full dynamic case and attempt to incorporate (11.5) into the dynamic model (11.1).

11.3.2 A Dynamic Model with Hysteresis

Up to this point, none of the constitutive models used in the dynamic case were able to capture the hysteretic behavior of elastomers. After developing the stress–strain law (11.6) with quasi-static data which successfully predicts hysteresis in several types of elastomers, we are ready to proceed in including (11.6) in the full dynamic model. This, however, leads to some immediate theoretical questions.

11.3.2.1 Theoretical Issues

By combining the constitutive relation (11.5) with (11.1), we obtain the following dynamic model for $t_K < t < t_{K+1}$:

$$
\rho A_c \frac{\partial^2 u}{\partial t^2} = \frac{\partial}{\partial x}\left[A_c \frac{\partial u}{\partial x} + A_c g_e\left(\frac{\partial u}{\partial x}, \frac{\partial^2 u}{\partial x \partial t}\right) + A_c C_D \frac{\partial^2 u}{\partial t \partial x} \right.
$$

$$
+ A_c Y(0) g_v\left(\frac{\partial u}{\partial x}\right) + A_c \int_{t-r}^{t} \dot{Y}(t-s) g_v\left(\frac{\partial u}{\partial x}, \frac{\partial^2 u}{\partial t \partial x}\right) ds
$$

$$
+ A_c \sum_{k=1}^{K} Y(t-t_k)(-1)^{k+1}\left[g_{vi}\left(\frac{\partial u}{\partial x}(t_k)\right) \right.
$$

$$
\left. \left. - g_{vd}\left(\frac{\partial u}{\partial x}(t_k)\right) \right] \right], \quad 0 < x < \ell,
$$

$$
A_c \left[\frac{\partial u}{\partial x} + g_e\left(\frac{\partial u}{\partial t}(t)\right) + C_D \frac{\partial^2 u}{\partial t \partial x} + Y(0) g_v\left(\frac{\partial u}{\partial x}(t), \frac{\partial^2 u}{\partial t \partial x}(t)\right) \right.
$$

$$
+ \int_{t-r}^{t} \dot{Y}(t-s) g_v\left(\frac{\partial u}{\partial x}(s), \frac{\partial^2 u}{\partial t \partial x}(s)\right) ds
$$

$$
+ \sum_{k=1}^{K} Y(t-t_k)(-1)^{k+1}\left[g_{vi}\left(\frac{\partial u}{\partial x}(t_k)\right) - g_{vd}\left(\frac{\partial u}{\partial x}(t_k)\right) \right] \right]\bigg|_{x=\ell} = f(t),
$$

$$
\tag{11.9}
$$

$$
u(0,x) = \varphi_0, \tag{11.10}
$$

$$
u_t(0,x) = 0, \tag{11.11}
$$

$$
u(t,0) = 0, \tag{11.12}
$$

$$
u(t,x) = \varphi_1, \quad t < 0. \tag{11.13}
$$

Here we assume that the elastomer, with the usual fixed end at $x = 0$, is at rest at time $t = 0$ with deformation φ_0 and memory φ_1. In addition, we include an (internal) Kelvin–Voigt damping term $A_c C_D u_{xxt}$ $(C_D > 0)$ in the model.

At this stage, it is unclear whether the above PDE is in fact well-posed. That is, can we even guarantee the existence of a solution for (11.3.2.1)–(11.13)? This issue is fully addressed in [4]; we present an outline of the results here. We note that a complete treatment of well-posedness questions requires functional analysis arguments that involve standard graduate-level mathematical concepts.

Consider the Hilbert spaces $V = H_L^1(0, \ell) = \{\phi : \phi \in H^1(0, \ell), \phi(0) = 0\}$ and $H = L^2(0, \ell)$, so that we have the Gelfand triple $V \hookrightarrow H \hookrightarrow V^*$. We denote by $\langle \cdot, \cdot \rangle$ the inner product in H, while $\langle \cdot, \cdot \rangle_{V^*,V}$ stands for the usual duality product (see [8], [29] for basic definitions and background). Let $\mathcal{L}_{[a,b]}$ denote the space of functions $u : [a, b] \to H$ such that

$$\mathcal{L}_{[a,b]} = \{u : u \in L^2(a, b; V), \ u_t \in L^2(a, b; V)\}.$$

Motivated by the model (11.3.2.1)–(11.13), we consider the system

$$
u_{tt} - C_D u_{xxt} - \left(u_x + g_e(u_x) + Y(0)g_v(u_x) + \int_{t-r}^t \dot{Y}(t-s)g_v(u_x(s))ds \right.
$$
$$
\left. + \sum_{k=1}^K Y(t-t_k)(-1)^{k+1}[g_{vi}(u_x(t_k)) - g_{vd}(u_x(t_k))] \right)_x
$$
$$
= F(t) \text{ in } V^*,
$$
$$(11.14)$$

$$u(t, 0) = 0, \tag{11.15}$$

$$u(a, x) = \varphi_0, \tag{11.16}$$

$$u_t(a, x) = 0, \tag{11.17}$$

$$u(t, x) = \varphi_1, \ t < a, \tag{11.18}$$

for $a \leq t_K \leq t \leq t_{K+1} = b$.

We say that $u \in \mathcal{L}_{[a,b]}$ is a weak solution of (11.14)–(11.18) with $\varphi_0 \in V$ and $\varphi_1 \in L^2(a - r, a; V)$ if it satisfies

$$
\int_a^t \left[-\langle u_t, \varphi_t \rangle + \langle u_x, \varphi_x \rangle + C_D \langle u_{xt}, \varphi_x \rangle \right.
$$
$$
+ \langle Y(0)g_v(u_x, \dot{u}_x), \varphi_x \rangle + \langle g_e(u_x), \varphi_x \rangle
$$
$$
+ \left\langle \sum_{k=1}^K Y(\tau - t_k)(-1)^{k+1}[g_{vi}(u_x(t_k)) - g_{vd}(u_x(t_k))], \varphi_x \right\rangle
$$
$$
+ \left. \left\langle \int_{\tau-r}^\tau \dot{Y}(\tau - s)g_v(u_x, \dot{u}_x)ds, \varphi_x \right\rangle \right] d\tau + \langle u_t(t), \varphi \rangle
$$
$$
= \int_a^t \langle F, \varphi \rangle_{V^*,V} d\tau \tag{11.19}
$$

for any $a \leq t_K \leq t \leq t_{K+1} = b$ and $\varphi \in \mathcal{L}_{[a,b]}$ as well as the initial condition

$$u(a, x) = \varphi_0. \tag{11.20}$$

Here we assume that g_e, g_{vi}, and g_{vd} are continuous nonlinear mappings of real gradient type, so that there exist continuous Frechet-differentiable nonlinear functionals which obey a certain type of boundedness. Moreover, the nonlinear functions g_e, g_{vi}, and g_{vd} are bounded in a particular sense, and their Frechet derivatives are bounded in $\mathcal{L}(H, H)$. We also assume that g_e, g_{vi}, and g_{vd} obey a monotonicity condition, and that Y is a smooth and bounded function. For details and more complete statements of these assumptions, see [4].

It is shown in [4] that under the precise assumptions detailed there, there exists a global weak solution of (11.14)–(11.18) for any $\varphi_0 \in V$ and $\varphi_1 \in L^2(a-r, a; V)$. We first proved that, under these assumptions, a weak solution $u^{(1)} \in \mathcal{L}_{[0, t_1]}$ exists for any $\varphi_0 \in V$ and $\varphi_1 \in L^2(-r, 0; V)$, i.e. we used $a = t_0 = 0$, $b = t_1$ in our definition of the weak solution. We showed that $u^{(1)}$ is smooth enough so that $u^{(1)}(t_1)$ and $(\partial u^{(1)}/\partial x)(t_1)$ exist. Next we suppose that a unique weak solution $u^{(K)}(t)$ exists on $[0, t_K]$ and consider the interval $[t_K, t_{K+1}]$. Then we have a similar system as above except that now we pick up a jump term, and φ_0 and φ_1 are modified, i.e. $\tilde{\varphi}_0 = u^{(K)}(t_K)$ and

$$\tilde{\varphi}_1 = \begin{cases} \varphi_1 & \text{if } t < 0, \\ u^{(K)} & \text{if } 0 < t < t_K. \end{cases}$$

We again show that a unique weak solution $\tilde{u}$ exists on this interval with the necessary smoothness, and

$$u^{(K+1)}(t) = \begin{cases} u^{(K)} & \text{if } t \leq t_K, \\ \tilde{u} & \text{if } t_K < t < t_{K+1}. \end{cases}$$

is a weak solution on $[0, t_{K+1}]$.

To show the existence of the weak solution $u^{(1)}$ on $[0, t_1]$, we first give *a priori* estimates, and then introduce Galerkin approximations to the weak solution and justify taking the limit. We used the Minty–Browder technique in showing that the limit of the Galerkin approximates is a weak solution. A complete statement of the theorem and its proof can be found in [4].

11.3.2.2 Numerical Results

After addressing the theoretical issues associated with the dynamic hysteretic model, we tested the effectiveness of this model on our previously obtained dynamic elastomer data. We used the free-release dynamic experiments with a 3-pound tip mass attached to the free end of a lightly filled carbon black sample. To model this particular sample, we use a cubic polynomial for $g_e(\epsilon) = E_1\epsilon + E_2\epsilon^2 + E_3\epsilon^3$ (as in the quasi-static case), and we suppose that g_v is linear and does not depend on $\dot{\epsilon}$, i.e. $g_v = g_{vi} = g_{vd}$. As before, we used $Y(t) = C_2 e^{-C_1 t}$, where $C_1, C_2 > 0$.

We note that although our dynamic model (11.3.2.1) contains Kelvin–Voigt damping, we do not include it in our computational model. This damping term is important for the theoretical result, but in our experiments and computations, we have determined that the hysteresis portion of the model can provide adequate damping to capture the dynamic behavior. We therefore can take the Kelvin–Voigt coefficient as $C_D \approx 0$, and thus reconcile the theoretical and numerical results.

Using the load cell data from the experiment, we set up a parameter identification problem to estimate the parameters $\rho, E_1, E_2, E_3, C_1, C_2$ (which we will denote collectively as q). That is, we wish to minimize

$$J(q) = \sum_{i=1}^{M} \left| z_i - A_c \sigma \left(\frac{\partial u}{\partial x} (t_i, 0; q) \right) \right|^2,$$

where z_i $(i = 1, \ldots, M)$, are observations from the load cell. Moreover, u is a solution to (11.3.2.1) with parameters q.

In our computations, we used linear splines for spatial discretization, while piecewise constant elements were used in the hysteretic discretization. (More details on the computational technique for dealing with the integral term can be found in [1], [2].) We used MATLAB optimization routines for the inverse problem. The computed result shows very good agreement with the collected data, with a relative error of 3% (see figure 11.7). Current efforts involve use of these models and techniques with experimental data from highly filled samples.

11.4 Conclusion

In the above presentation, we have outlined the collaborative effort between applied mathematicians in the CRSC and scientists at Lord Corporation, focused on the modelling of elastomers. We have made significant progress to date towards the final goal of developing a high-quality model for the dynamic behavior of elastomers.

The initial phase of the project was an exploration of nonlinear constitutive laws. We considered several types of nonlinear stress–strain law, including many found in the existing literature. Working with both the quasi-static and the dynamic cases, our experiments and computations suggested that neither the neo-Hookean nor a linear constitutive relation are adequate for capturing the dynamic behavior of elastomers, even for unfilled natural rubber. For very

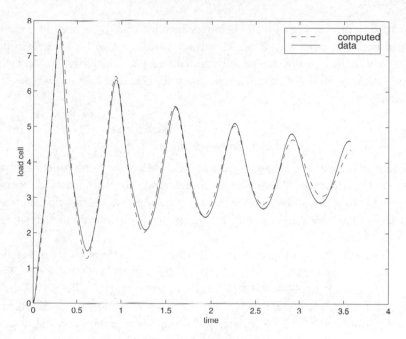

Fig. 11.7. Dynamic results for rubber lightly filled with carbon black

lightly filled and unfilled rubber, it appears that a nonlinear stress–strain law is sufficient, while for more highly filled rubbers it is clear that hysteresis must be included in the constitutive model.

The next phase of our collaboration involved incorporating hysteresis into our quasi-static and dynamic models. We refined our stress–strain law first using quasi-static data, until we obtained a model that could accurately predict the quasi-static behavior of several types of elastomer. We then returned to the dynamic case to incorporate the hysteretic constitutive relation in our dynamic model. After formulating the full dynamic PDE model with hysteresis, it became clear that the question of the well-posedness of the model needed to be resolved in a formal manner. This led to the development of the weak form of the PDE and a proof of the existence of solutions. Once these theoretical questions were addressed, we tested the accuracy of this model in the dynamic case by using inverse problem techniques. As shown in the previous section, this model produces satisfactory results for lightly filled rubber. Current work includes the testing of our dynamic model on more highly filled rubbers, which exhibit more significant hysteresis.

Future work in this collaboration includes a study of the mechanical behavior of elastomers in shear. Since the overall goal of the project is to accurately predict the dynamic behavior of elastomers, we need to understand both extension and shear. As with extension, we will work with both the quasi-static and dynamic paradigms to develop a model for shear, and we will test the accuracy of our model by gathering shear data for various types of elastomer.

Acknowledgments

We are grateful to Yue Zhang (current address: Michelin North America) and N.J. Lybeck (current address: Applied Math, Inc.) for their substantial contributions to many of the earlier efforts described above. We would also like to thank Richard Wilder of Lord Corporation for his help with implementing the experiments.

This research was supported in part by the U.S. Air Force Office of Scientific Research under grants AFOSR F49620-95-1-0236, AFOSR F49620-95-1-0375, AFOSR F49620-98-1-0180, and in part by an NSF-GRT fellowship to L.K.P. under grant GER-9454175 and a GAANN fellowship to Y. Zhang under grant P200A40730. N. Lybeck was supported in part by an NSF University/Industrial postdoctoral grant NSF DMS 9508617. Graduate students and postdoctorates also received generous support (both intellectual and financial) from the Lord Corporation.

References

[1]	Banks, H. T., Fabiano, R. H. & Wang, Y. (1988) Estimation of Boltzmann damping coefficients in beam models. In: *COMCON Conf. on Stabilization of Flexible Structures* (ed. Balakrishnan, A. V.) New York, USA, 13–35.

[2]	Banks, H. T., Fabiano, R. H. & Wang, Y. (1989) Inverse problem techniques for beams with tip body and time hysteresis damping. Matemática Aplicada e Computacional **8**, 101–118.

[3]	Banks, H. T., Lybeck, N. J., Gaitens, M. J., Muñoz, B. C. & Yanyo, L. C. (1996) Computational methods for estimation in the modeling of nonlinear elastomers. Kybernetika **32**, 526–542.

[4]	Banks, H. T., Pintér, G. A. & Potter, L. K. (2000) Existence of unique weak solutions to a dynamical system for nonlinear elastomers with hysteresis. Differential Integral Equations, **13**, 1001–1024.

[5]	Banks, H. T., Pintér, G. A., Potter, L. K., Gaitens, M.J. & Yanyo, L.C. (1999) Modeling of nonlinear hysteresis in elastomers under uniaxial tension. J. Intel. Mat. Syst. Str. **10**, 116–134.

[6]	Banks, H. T., Pintér, G. A., Potter, L. K., Muñoz, B. C. & Yanyo, L. C. (1998) Estimation and control related issues in smart material structures and fluids. In *4th International Conference on Optimization Techniques and Applications* (eds. Caccetta, L. et al.) Perth, Australia, 19–34.

[7] Banks, H. T., Potter, L. K. & Zhang, Y. (1996) Stress-strain laws for carbon black and silicon filled elastomers. In *Proc. 36th IEEE Conf. on Dec. and Control*, 3727–3732.

[8] Banks, H. T., Smith, R. C. & Wang, Y. (1996) *Smart Material Structures: Modeling, Estimation and Control*, Masson/Wiley, Paris/Chichester.

[9] Bloch, R., Chang, W. V. & Tschoegl, N. W. (1978) The behavior of rubberlike materials in moderately large deformations. Journal of Rheology **22**, 1–32.

[10] Charlton, D. J., Yang, J., & Teh, K. K. (1994) Mechanical properties of solid polymers. Rubber Chemistry & Technology **67**, 481–503.

[11] Christensen, R. M. (1982) *Theory of Viscoelasticity: An Introduction*, 2nd ed., Academic Press, New York.

[12] Clough, R. W. & Penzien, J. (1975) *Dynamics of Structures*, McGraw-Hill, New York.

[13] Crawford, R. J. (1985) *Plastics and Rubbers: Engineering Design and Applications*, Mechanical Engineering Publications Ltd, London.

[14] Ferry, J. D. (1980) *Viscoelastic Properties of Polymers*, John Wiley & Sons, New York.

[15] Gent, A. N. (1992) *Engineering with Rubber: How to Design Rubber Components*, Hanser Publishers, New York.

[16] Johnson, A. R., Tessler, A. & Dambach, M. (1997) Dynamics of thick viscoelastic beams. J. Engr. Materials and Tech. **119**, 273–278.

[17] Jolly, M. R., Carlson, J. D. & Muñoz, B. C. (1996) A model of the behavior of magnetorheological materials. Smart Mater. Struct. **5**, 607–614.

[18] Jolly, M. R., Carlson, J. D., Muñoz, B. C. & Bullions, T. A. (1996) The magnetoviscoelastic response of elastomer composites consisting of ferrous particles embedded in a polymer matrix. Journal of Intelligent Material Systems and Structures **7**, 613–622.

[19] Lesieutre, G. A. (1994) Modeling frequency-dependent longitudinal dynamic behavior of linear viscoelastic long fiber components. J. Composite Materials **28**, 1770–1782.

[20] Lesieutre, G. A. & Govindswamy, K. (1996) Finite element modeling of frequency-dependent and temperature-dependent dynamic behavior of viscoelastic materials in simple shear. Int. J. Solids Structures **33**, 419–432.

[21] Marsden, J. E. & Hughes, T. J. R. (1983) *Mathematical Foundations of Elasticity*, Prentice-Hall, Englewood Cliffs, NJ.

[22] Nagdi, K. (1993) *Rubber as an Engineering Material: Guideline for Users*, Hanser Publishers, New York.

[23] Ogden, R. W. (1984) *Non-Linear Elastic Deformations*, Ellis Horwood Limited, Chichester.

[24] Rivlin, R. S. (1948) Large elastic deformations of isotropic materials, I, II, III. Phil. Trans. Roy. Soc. A **240**, 459–525.

[25] Schapery, R. A. (1969) On the characterization of nonlinear viscoelastic materials. Polymer Engineering and Science **9**, 295–310.

[26] Timoshenko, S., Young, D. H., & Weaver, Jr., W. (1974) *Vibration Problems in Engineering*, J. Wiley & Sons, New York.

[27] Treloar, L. R. G. (1975) *The Physics of Rubber Elasticity*, Clarendon Press, Oxford, 3rd ed.

[28] Ward, I. M. (1983) *Mechanical Properties of Solid Polymers,* John Wiley & Sons, New York, 2nd ed.

[29] Wloka, J. (1992) *Partial Differential Equations,* Cambridge Univ. Press, England.

[30] Zhang, Y. (1997) *Mathematical formulation of vibrations of a composite curved beam structure: Aluminum core material with viscoelastic layers, constraining layers and piezoceramic patches*, Ph.D. Thesis, N.C. State University.

H. T. Banks, Gabriella A. Pintér, and Laura K. Potter
Center for Research in Scientific Computation,
North Carolina State University, Raleigh, NC, USA
htbanks@eos.ncsu.edu
gapinter@unity.ncsu.edu
lkpotter@unity.ncsu.edu

Michael J. Gaitens and Lynn C. Yanyo
Thomas Lord Research Center, Lord Corporation, Cary, NC, USA
info@lordcorp.com
info@lordcorp.com

12

A Gasdynamic–Acoustic Model of a Bird

Scare Gun

12.1 Introduction

A bird scare gun is a relatively simple device which produces impulsive noise by means of a gas explosion. It is used for scaring birds away from areas where their presence is unwanted, like orchards, airfields, or oilfields. It is meant to operate automatically for long periods of time, with little or no human intervention. The construction is simple and robust, with as few as possible moving parts, so that a lifecycle in the order of 100,000–200,000 explosions is attainable.

The mechanism is simple. A carefully controlled mixture of air and propane or butane gas (stoichiometric† mixture, or a little bit richer than that) is periodically (every 5 or 10 minutes) blown into a semi-open pot, which is the combustion chamber. This pot is connected via a small diaphragm or iris (a small hole in the wall of the combustion pot) to an exhaust pipe. After ignition, the gas burns quickly (but without detonation, i.e. with a subsonically moving flame front) so that pressure and temperature increase quickly. This high pressure drives the gas out of the pot via a hot jet, which issues from the diaphragm into the pipe. Acting like a piston, this jet creates a pressure wave in the cold exhaust pipe. Part of the wave reflects at the exit, and part radiates, nearly spherically, away into the open air.

An interesting detail in the design gives, without any further analysis, insight in the gasdynamic behavior. In order to vary the noise that is produced, the length of the exhaust pipe is made variable. The pipe consists of two shorter pipes, one of which slides inside the other, like a telescope. It appears that the gun produces a louder bang when the pipe is longer. This hints, of course, at a possible relation to nonlinear wave steepening. A pressure wave of high-enough amplitude steepens while propagating onto the pipe exit. Indeed, from fairly elementary acoustics, it is known that (at least for low-frequency linear perturbations) the radiated

---

† A stoichiometric mixture has just enough of each component for a complete chemical reaction.

part is proportional to the spatial derivative of the pipe wave. Therefore, a longer pipe, leading to more advanced steepening, may be expected to produce more noise. So the effect of steepening is not only to intensify the higher frequencies – the pitch – of the radiated pulse but also to increase its amplitude.

The size of the diaphragm is also known to be very critical for the loudness of the sound. On one hand, too large a hole would allow the gas to be blown away too quickly so that not all the gas reacts, and the pressure wave remains too flat. On the other hand, with too small a hole the mass flux of the jet will remain too small, and hence the created pressure wave will be too small. Somewhere there is obviously an optimum.

At present the design of a bird scare gun is mainly done by trial and error using the skills of experienced craftsmen. A mathematical model would enable the designer to understand the various physical and chemical processes better, and hence to further refine the quality of the gun.

Aspects of practical interest are, for example:

- How much chemical energy is needed to produce a bang of a given loudness?
- How does the length of the exhaust pipe relate to the noise that is produced?
- How do we determine the optimum iris?
- What are the maximum pressure and temperature levels in the pot? (Although it is difficult to quantify the relation, the life of the device appears to be determined completely by wear due to fatigue of the pot.)

Most of these questions cannot fully be answered by a mathematical model of manageable complexity. Three-dimensional combustion, flame front propagation, turbulent compressible hot–cold gas mixing, etc. are complex phenomena, which are very difficult to model, while any model based on first principles will be very difficult to evaluate. So we will try to proceed, as far as possible, with the answers to the above design questions as ideal goals.

To investigate the effect of various problem parameters on the resulting noise level (sound pressure level, SPL), we will develop a model that is simple enough to make progress, but still rich enough to describe the essence of the underlying mechanisms. As we will see, the problem is physically quite complicated, and requires many modeling assumptions. A great help in this respect is the fact that noise is measured logarithmically (in decibels), which turns rather crude assumptions into relatively accurate results.

Fig. 12.1. Sketch of geometry

12.2 Model

12.2.1 Geometry

A sketch of the device is shown in figure 12.1. In the pot (1) a mixture of butane or propane gas and air is ignited. From the pressure difference between (2) and (1) a hot jet (j) is formed from the iris into the cold pipe. The flux of mass and momentum across a hot–cold mixing zone at $x = 0$ initiates a pressure wave in the pipe. This wave steepens and reflects at the open end ($x = L$), where an almost spherical† sound pulse is generated that radiates away to the far field.

12.2.2 Pot

For a small enough iris, any flow in the pot is negligible. Also, the size of the pot seems to be just small enough to allow us to model the gas in the pot using mean quantities. We have a volume Ω of gas of density ρ_1, pressure p_1, and temperature T_1. The jet has cross section S_j, (average) velocity v_j, and density ρ_j, so conservation of mass requires that

$$\Omega \frac{d\rho_1}{dt} = -\rho_j v_j S_j. \tag{12.1}$$

The jet cross section S_j is smaller than the diaphragm S_d because of the vena-contracta effect (radial inertia narrowing the jet). Typically, S_j/S_d is of the order 0.6, but the actual value depends on various parameters like the Reynolds number, the Mach number, and the sharpness of the iris edges [4].

The typical duration of pressure build up and release is a few milliseconds, which corresponds to a pressure wave length of several decimeters.

---

† Well, not exactly. In general the situation may be very complex. The higher-frequency components tend to propagate straight out of the exit like rays, whereas the lower frequencies spread out spherically. Furthermore, high-amplitude waves lose acoustic energy by vortex shedding at the sharp exit edges. Measurements indicate a variation of a few dB between the forward and backward sound fields. Therefore we start with the simplest assumption of spherical radiation.

(The speed of sound is of the order of 34 cm/ms.) This is large compared to the iris and jet size. On the other hand, the jet is small compared to the pipe length. So the jet may be described as quasi-stationary and local. Furthermore, a typical jet velocity is high (nearly sonic), so viscous effects are small, and the flow from pot to jet is compressible. Assuming a nearly optimal situation, such that there is no loss of mechanical energy by shock formation, the flow is isentropic, for which Bernoulli's law and the condition of isentropy are valid:

$$\frac{\gamma}{\gamma - 1}\frac{p_1}{\rho_1} = \frac{1}{2}v_j^2 + \frac{\gamma}{\gamma - 1}\frac{p_j}{\rho_j}, \tag{12.2}$$

$$\frac{p_1}{\rho_1^\gamma} = \frac{p_j}{\rho_j^\gamma}. \tag{12.3}$$

The ideal gas law and the definition of sound speed c are valid everywhere (pot, pipe, outside):

$$p = \rho R T, \tag{12.4}$$

$$c^2 = \left(\frac{\partial p}{\partial \rho}\right)_S = \frac{\gamma p}{\rho} = \gamma R T. \tag{12.5}$$

The heat capacity at constant pressure, $C_P = 1000.0$ J/kg K; the heat capacity at constant volume, $C_V = 713.26$ J/kg K; their ratio $\gamma = C_P/C_V = 1.402$; and their difference, the gas constant $R = C_P - C_V = 286.73$ J/kg K are taken to be constant; although in reality they vary slightly with temperature.

The combustion appears to be very difficult to model from first principles. The finite duration is acoustically important, so a description of the propagating flame front seems to be required. This is very complicated if the effects on temperature and pressure of a decreasing density due to the outflow from the iris are to be included.

To make progress, some experimental input is used at this point. The effect of the increase of thermal energy from the combustion is described by an increase of the entropy, in the form of a suitably chosen function $\beta(t)$

$$\beta(t) = \beta_0 + (\beta_1 - \beta_0)b(t/\tau) = \frac{p_1}{\rho_1^\gamma}. \tag{12.6}$$

Guided by experiments (figure 12.2), the following shape appeared to be realistic (see figure 12.3)

$$b(\zeta) = \begin{cases} 0 & \zeta \in (-\infty, 0.5], \\ \dfrac{37\exp(4-(\zeta-2.5)^2)/(1+9(\zeta-2.5)^2)-1}{37\exp(4)-1} & \zeta \in [0.5, 2.5], \\ 1 & \zeta \in [2.5, \infty). \end{cases}$$

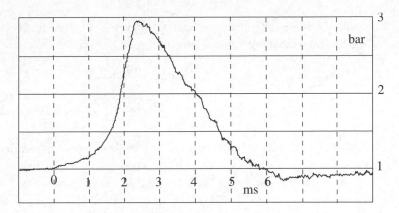

Fig. 12.2. Measured pressure inside the pot

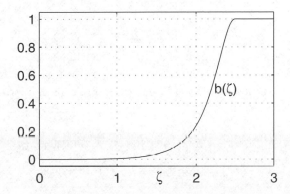

Fig. 12.3. Shape function for the entropy increase

The initial and final levels β_0 and β_1 and the combustion time τ are to be found from experiments. The parameter β_0 may be found directly from p_∞ and ρ_∞. If only the pressure in the pot can be measured, and not the temperature or the density, β_1 can be found by trial and error, until the maximum pot pressure fits the measured value. The detailed description of $b(\zeta)$ is of minor importance.

An estimate for the increase in entropy may now be found as follows. In the *theoretical* case of a completely closed pot, so that $\rho_1 = \rho_\infty = $ constant, the variation in entropy ds and internal energy de are given by the fundamental law of thermodynamics for reversible processes, and the relation for a perfect gas

$$T ds = de + p d\rho^{-1} = de = C_V dT = \frac{C_V}{\rho_\infty R} dp.$$

So the energy increase ΔQ_1, and the entropy increase Δs_1, are functions only of the initial and theoretical final temperature or pressure:

$$\Delta Q_1 = \int T ds = C_V (T_{th} - T_\infty) = \frac{p_{th} - p_\infty}{\rho_\infty (\gamma - 1)},$$

$$\Delta s_1 = \beta_1 - \beta_0 = \frac{p_{th} - p_\infty}{\rho_\infty^\gamma}.$$

It follows that, if we knew the total amount of chemical energy transferred into heat, and if the pot were closed, then we would know the increase of entropy. In practice this is an upper limit, because the pot is open: we do not know exactly the amount of gas that reacts, since a part of it is blown away into the pipe, and an a priori unknown part of the energy increase $(p d \rho^{-1})$ is used for a volume change.

12.2.3 Jet

When the jet becomes turbulent, the flow is decelerated, and the jet widens until it fills the whole cross section S_2 of the pipe. The transition from the jet (j) to the pipe (2) is described by crosswise averaged conservation laws [1, 4]:

$$S_j \rho_j v_j = S_2 \rho_2 v_2 \qquad \text{(mass)}, \qquad (12.7)$$

$$S_2 p_j + S_j \rho_j v_j^2 = S_2 p_2 + S_2 \rho_2 v_2^2 \qquad \text{(momentum)}, \qquad (12.8)$$

$$S_j \rho_j v_j \left(\tfrac{1}{2} v_j^2 + w_j\right) = S_2 \rho_2 v_2 \left(\tfrac{1}{2} v_2^2 + w_2\right) \quad \text{(energy)}, \qquad (12.9)$$

where $w = e + p/\rho$ denotes the heat function [2]. It should be noted that the pressure p_j inside the jet is equal to the value outside. For a perfect gas, the heat function is given by [2, p.315] $w = (\gamma/\gamma - 1)p/\rho$, and the energy integral may be further simplified to

$$\frac{1}{2} v_j^2 + \frac{\gamma}{\gamma - 1} \frac{p_j}{\rho_j} = \frac{1}{2} v_2^2 + \frac{\gamma}{\gamma - 1} \frac{p_2}{\rho_2}. \qquad (12.10)$$

Note that this equation is similar to Bernoulli's equation (the integral of the momentum equation along a streamline). However, Bernoulli is not relevant here, because the flow is not isentropic. Energy is lost in the form of conversion into turbulence (which has not been modeled).

At a later stage in the process, the pressure in the pipe may be increased by the reflected wave, and the pot pressure may become lower than the pipe pressure, leading to a jet flow *into* the pot. Assuming the iris to behave symmetrically, the equations are the same as above, apart from a sign in the velocity. Note that this inward jet forms a sink of acoustic energy.

12.2.4 Pipe

When the hot jet enters the cold pipe, a hot–cold mixing zone is created. This is gradually convected into the pipe, and describing this process calls again for serious modeling assumptions.

Since the jet velocity v_j is at most sonic, and since the iris S_d is much smaller than the pipe cross section S_2, the pipe velocity v_2 is much smaller than the sound speed, and the propagation of the mixing zone can be ignored compared to the generated pressure wave. Thus only the flux of momentum across the mixing zone needs to be taken into account.

Since the mixing zone is small, the flux of momentum can be described in a quasi-stationary manner by continuity of pressure. Together with continuity of mass flux, we have the following relations between the values at the hot side (subscript 2) and the cold side (subscript 0, referring to $x = 0$):

$$p_2 = p_0, \tag{12.11}$$

$$p_2 v_2 = p_0 v_0. \tag{12.12}$$

Note that we defined $x = 0$ as the position of the mixing zone between the hot jet and the cold pipe gas. This position is difficult to identify, but on the other hand, the jet is relatively small. Therefore we take $x = 0$ simply at the position of the iris.

Inside the pipe, we assume that the generated pressure wave (and its reflections) are of linear acoustic or weakly nonlinear type, such that the temperature and sound speed are approximately constant, but not exactly (no isothermal changes). So we cannot put p_0 equal to $\rho_0 R T_\infty$.

Let us start by assuming that the amplitude of the wave in the pipe is small enough for an acoustic (linear, isentropic) one-dimensional approximation to be valid. The gasdynamic equations are then

$$\rho_t + \rho_\infty v_x = 0, \quad \rho_\infty v_t + p_x = 0, \quad dp = c_\infty^2 d\rho,$$

where the indices x and t indicate partial derivatives with respect to x and t. The values at $x = 0$ are as yet unknown, as they are coupled to the jet flow, but at $x = L$ the flow leaves the exit nearly spherically (at least for small pipe diameters). The outside velocity drops very quickly, so that the pressure near the exit plane is practically equal to the ambient pressure. So we take the boundary condition of a vanishing acoustic pressure

$$p(L, t) - p_\infty = 0.$$

The effective position of the open end (as seen by the wave incident from the pipe) is slightly outside the exit plane. Typically, for low frequencies and a thin pipe wall, it is located at $x = L + 0.6a$, where a is the pipe radius, and $0.6a$ is called the "end correction". The error made by ignoring the end correction compensates to some extent the above error made in the position $x = 0$.

When we eliminate p, ρ, or v from the above linearised equations, we obtain the 1-D wave equation

$$p_{tt} - c_\infty^2 p_{xx} = 0,$$

which has the well known solution of d'Alembert [2, p.246]. This leads, when combined with the boundary condition at $x = L$, to

$$p(x, t) = p_\infty + \rho_\infty c_\infty^2 \left[f\left(t - \frac{x}{c_\infty}\right) - f\left(t + \frac{x - 2L}{c_\infty}\right)\right], \qquad (12.13)$$

$$\rho(x, t) = \rho_\infty + \rho_\infty \left[f\left(t - \frac{x}{c_\infty}\right) - f\left(t + \frac{x - 2L}{c_\infty}\right)\right], \qquad (12.14)$$

$$v(x, t) = c_\infty \left[f\left(t - \frac{x}{c_\infty}\right) + f\left(t + \frac{x - 2L}{c_\infty}\right)\right], \qquad (12.15)$$

with a shape function f that is to be determined. Since we have silence for all $x \in [0, L]$ when $t < 0$, f satisfies the condition of causality:

$$f(z) = 0 \quad \text{for } z \le 0. \qquad (12.16)$$

One should be aware of the fact that the exit boundary condition of vanishing acoustic pressure is quite sufficient to determine roughly the amplitude of the reflected wave, but in other respects is very crude. For example, it leads to equal amplitudes of the incident and reflected waves, which means that as much energy is incident as is reflected. In other words, there is no energy loss at the exit, and the wave would bounce back and forth for ever if there was no vortex shedding from the iris.

In practice there is energy loss by radiation and vortex shedding, all of which are rather difficult phenomena to model. Therefore, it may be better to reduce the reflected amplitude a little, by writing $Rf(t + (x - 2L/c_\infty))$, where R is a number slightly smaller than 1. The problem here, of course, is that we have little information on the appropriate value to be taken (especially because it is an ad-hoc "fix", and is not based on any modelling of the physics). For this reason, we will not pursue the matter further here.

12.2.5 Radiated Field

Outside the pipe, we have relatively low amplitudes of the perturbation $p - p_\infty$, and the gasdynamic problem certainly simplifies to an acoustic one, with governing equation

$$p_{tt} - c_\infty^2 \nabla^2 p = c_\infty^2 Q_L'(t)\delta(x)$$

where $Q_L(t)$ is the mass flux from the pipe exit, while the coordinate system has been changed a little, so that the origin is now taken to be at the exit plane. From symmetry, the field depends only on t and r, where $|x| = r$. The solution (for outgoing waves) is [2, p.265]

$$p(r, t) = p_\infty + \frac{1}{4\pi r} Q_L'\left(t - \frac{r}{c_\infty}\right).$$

Since the mass flux is equal to

$$Q_L(t) = \rho_\infty S_2 v(L, t) = 2S_2 \rho_\infty c_\infty f\left(t - \frac{L}{c_\infty}\right), \tag{12.17}$$

we have for the outer field

$$p(r, t) = p_\infty + \frac{\rho_\infty c_\infty S_2}{2\pi r} f'\left(t - \frac{L}{c_\infty}\right), \tag{12.18}$$

which shows that the radiated field is proportional to the derivative of the pipe field.

Finally, we wish to assign a single number to the sound field, in order to measure its loudness. This is not straightforward, because the time history of the pressure is a continuous function, which may be interpreted as an infinite-dimensional vector. (Or, if sampled, a vector of finite but high dimension.) The problem is to find a norm, suitably fitting with the physiology of our and birds' ears and brains. Spectral distribution, duration, presence of pure tones, and many more factors may play a role, and as a result there exist many definitions of noise level.

Since the range of our and birds' audial sensitivity is incredibly large (10^{14} in energy), the loudest and quietest levels are almost always practically infinitely far away, and we have no reference, or scaling level, to compare with, other than the sound itself that we are hearing. As a result, sound intensity is (necessarily) perceived logarithmically. Therefore, the various definitions of sound level always include a (base-10) logarithm.

One of the most elementary and widely used definitions is the SPL (short for sound pressure level). It is based on the acoustic energy density, averaged over a short, but not too short period of time. Since acoustic energy is proportional to $(p - p_\infty)^2$, this averaged energy is (apart from a square) equivalent to the root-mean-square (rms) of $p - p_\infty$. This averaged acoustic pressure is then compared to a reference level of $2 \cdot 10^{-5}$ Pa, which corresponds (for good ears and optimal conditions) to the threshold of hearing.

By taking the logarithm of the ratio $(\text{rms}/2 \cdot 10^{-5})^2$ we obtain the level in bels, and hence in decibels if we multiply this by 10.

The noise level (sound pressure level, SPL) at a distance r is thus defined by

$$\text{SPL} = 20 \log_{10} \left[\left\{ \frac{1}{T_{\text{ref}}} \int_0^{T_{\text{ref}}} (p(r, t) - p_\infty)^2 \, dt \right\}^{1/2} /2 \cdot 10^{-5} \right] \text{dB},$$

$$(12.19)$$

where by definition $T_{\text{ref}} = 1$ s for "slow" measurements (noise of long duration), $T_{\text{ref}} = \frac{1}{8}$ s for "fast" measurements (short duration), and $T_{\text{ref}} = 35$ ms for "impulsive" measurements (impulsive sound). The present noise character is in general of impulsive type, since the main pressure variations (the part that contributes the most to the SPL value) occur within 35 ms. If the noise duration is less than 35 ms, the difference between these three measuring definitions is of course just 9 dB and 5.5 dB. On the other hand, if the problem parameters were changed so that the pot emptied much more slowly (for example by reducing the size of the iris), then the time history would extend much beyond the 35 ms, and it would be necessary to use another definition.

These formal definitions of SPL have absolutely no bearing on the physical phenomena, of course, but they indicate that the typical accuracy of any suitable model is in the order of decibels, which is rather crude from a gasdynamics point of view.

12.2.6 Nonlinear Correction in the Pipe

A very important observation that we can make from equation (12.18) concerns the fact that the outside pressure depends on the derivative f', rather than on f itself. So any steepening of the wave inside the pipe is immediately coupled to the outside noise level. Therefore, it seems relevant, and maybe necessary, to investigate here the possibility of incorporating an extension of the foregoing model in which we include the acoustic effects of the wave steepening.

The general problem of how the left and right running compression waves behave is too difficult for any analytical approach. The main effect here, however, is the creation of a wave generated by the jet, and incident to the pipe exit. Therefore, we consider a 1-D isentropic gas flow, with a perturbation propagating in the positive direction, given by

$$\rho_t + (\rho v)_x = 0, \quad \rho(v_t + v v_x) + p_x = 0, \quad dp = c^2 d\rho. \qquad (12.20)$$

If we assume the velocity and pressure to be functions of ρ only, and follow the disturbance along $x = X(t)$, we have [2, p.366]

$$\rho_t + \rho v' \rho_x + \rho_x v = 0, \quad \rho v' \rho_t + \rho v v' \rho_x + c^2 \rho_x = 0,$$

so that

$$\frac{d}{dt}\rho(X(t), t) = \rho_t + \dot{X}\rho_x = \rho_t + (\rho v' + v)\rho_x = \rho_t + \left(v + \frac{c^2}{\rho v'}\right)\rho_x = 0,$$

which is possible only if

$$\rho v' = \pm c,$$

leading to

$$X = X_0 + (v \pm c)t,$$

because ρ and hence v and p are constant along $x = X(t)$. Furthermore, isentropy implies that $2c^{-1}dc = (\gamma - 1)\rho^{-1}d\rho$, so that

$$\frac{dv}{dc} = \frac{dv}{d\rho}\frac{d\rho}{dc} = \pm\frac{2}{\gamma - 1}$$

which leads to

$$c = c_o \pm \tfrac{1}{2}(\gamma - 1)(v - v_0).$$

Since v_0 is usually much smaller than c_0, the development of a nonlinear acoustic perturbation, propagating from $x = x_0$ at $t = t_0$ to a point x at time t, is therefore described by [3, p.569]

$$x - x_0 = (c_0 + \tfrac{1}{2}(\gamma + 1)v)(t - t_0). \tag{12.21}$$

In other words, the perturbations to p_0 and v_0 that start at $x_0 = 0$ at time t_0 arrive at $x = L$ at time

$$t_L = t_0 + \frac{L}{c_0(t_0) + \tfrac{1}{2}(\gamma + 1)v_0(t_0)}. \tag{12.22}$$

We see that perturbations with a larger velocity $v_0(t_0)$ propagate faster (t_L is smaller). Figure 12.4 illustrates this phenomenon: in the first picture the excess density is plotted against x. The top of the wave travels faster than the bottom and so it steepens. In the second and third pictures the excess density is plotted against t. We observe that the wave passes through quicker at $x = x_2$ than at $x = x_1$.

When t is small enough, the wave propagates in the pipe only in the forward direction, so the present approach is certainly valid. When the backward-running reflected part is added, the amplitudes change and so do

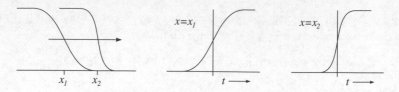

Fig. 12.4. Sketch of nonlinear wave steepening

the propagation speeds. We will, however, ignore this aspect of the problem and assume the forward and backward running waves to be uncoupled, because this stage of the process is already of less importance acoustically.

As a result, the acoustic field (12.13), (12.14), and (12.15) is modified such that f follows from

$$v(L, t) = 2c_\infty f\left(t - \frac{L}{C^+(\tilde{v}_0)}\right),\tag{12.23}$$

where

$$C^\pm(v) = c_\infty \pm \tfrac{1}{2}(\gamma + 1)v,$$

and $\tilde{v}_0$ is the value of v at $x = 0$ at a time $L/C^+(\tilde{v}_0)$ earlier. Furthermore,

$$v_0 = v(0, t) = c_\infty\left[f(t) + f\left(t - \frac{L}{C^+(\hat{v}_0)} - \frac{L}{C^-(\hat{v}_L)}\right)\right],\tag{12.24}$$

and $\hat{v}_0$ is the value of v at $x = 0$ at a time $L/C^+(\hat{v}_0) + L/C^-(\hat{v}_L)$ earlier, and $\hat{v}_L$ is the value of v at $x = L$ at a time $L/C^-(\hat{v}_L)$ earlier. Once we have v, similar expressions for p and ρ follow immediately.

This process is easily evaluated sequentially in time, because of the causality condition (12.16).

12.3 Analysis

To finally obtain results, we have to couple the pot (section 12.2.2) and the jet (section 12.2.3) with the pipe (sections 12.2.4 and 12.2.6) and the radiated field (section 12.2.5). By combining equations (12.2), (12.3), and (12.6) we can eliminate v_j to obtain

$$v_j = \left(\frac{2\gamma\beta}{\gamma - 1}\right)^{1/2}\left(\rho_1^{\gamma-1} - \rho_j^{\gamma-1}\right)^{1/2}.$$

Note that this cannot be valid if $\rho_1 < \rho_j$. Of course, this is because in that case the jet has changed direction and the flow is from pipe into the pot. This is a situation that only occurs in a rather late stage of the process, when the pot has emptied itself almost completely, and the reflected wave in the pipe temporarily creates a slightly higher pressure. Since this is acoustically rather unimportant, we will cure this problem by a simple measure and just change the flow direction, ignoring any further details of the flow. So we have

$$v_j = \text{sgn}(\rho_1 - \rho_j) \left(\frac{2\gamma\beta}{\gamma - 1} \right)^{1/2} \left| \rho_1^{\gamma-1} - \rho_j^{\gamma-1} \right|^{1/2}. \tag{12.25}$$

Thus v_j depends only on ρ_1 and ρ_j, and we have for equation (12.7) the following form:

$$\frac{d\rho_1}{dt} = F(\rho_1, \rho_j) = -\frac{S_j}{\Omega} \rho_j v_j, \tag{12.26}$$

which can be solved numerically for t as soon as we know the relationship between ρ_1 and ρ_j. From equations (12.7), (12.8), and (12.10) we can find expressions for $\rho_2 v_2$ and $\rho_2 v_2^2$:

$$\rho_2 v_2 = \sigma \rho_j v_j,$$

$$\rho_2 v_2^2 = \sigma \frac{2\gamma\beta}{\gamma - 1} (\rho_1^{\gamma-1} - \rho_j^{\gamma-1}) \rho_j + \beta \rho_j^{\gamma} - p_2 = \frac{2\gamma}{\gamma - 1} (\beta \rho_2 \rho_1^{\gamma-1} - p_2),$$

where $\sigma = S_j / S_2$.

Now we can eliminate ρ_2:

$$p_2 = \sigma \rho_j + \rho_1^{1-\gamma} \left(\frac{\gamma + 1}{2\gamma\beta} p_2 + \left(\frac{\gamma - 1}{2\gamma} - \sigma \right) \rho_j^{\gamma} \right). \tag{12.27}$$

For a given p_2, the relation $\rho_1 = \rho_1(\rho_j)$ is given by the algebraic equation

$$G(\rho_1, \rho_j) = \sigma \rho_1^{\gamma-1} \rho_j \left[\frac{p_2}{\gamma\beta} + \left(\frac{\gamma - 1}{\gamma} - \sigma \right) \rho_j^{\gamma} \right] + \left(\frac{\gamma - 1}{2\gamma} - \sigma \right)^2 \rho_j^{2\gamma}$$

$$\times \left(\frac{\gamma - 1}{2\gamma} - \sigma \right) \rho_j^{\gamma} \frac{p_2}{\gamma\beta} - \frac{1}{4} (\gamma^2 - 1) \left(\frac{p_2}{\gamma\beta} \right)^2 = 0. \tag{12.28}$$

To obtain for given ρ_1 the root ρ_j, we note that $G(\rho_1, 0) = -\frac{1}{4}(\gamma^2 - 1)$ $(p_2/\gamma\beta)^2 < 0$, and $G = O(\rho_j^{2\gamma}) > 0$ as $\rho_j \to \infty$, so there is at least one root. If $\sigma > 1 - \gamma^{-1}$ there may be more, in which case we need the smallest one. This one is always less than $(p_2/\gamma\beta)^{1/\gamma} (\sigma - 1 + \gamma^{-1})^{-1/\gamma}$, and there is only one.

Taking everything together, we have a differential equation (12.26) coupled with a system of algebraic equations ((12.25), (12.28), (12.6), (12.11), (12.12), (12.27)) and delay equations ((12.24), (12.23)) and analogous equations for p and ρ, which can be solved numerically, to yield the outside pressure (SPL) field (12.18) and (12.19).

We have combined a standard root-finding routine for $G(\rho_1, \rho_j) = 0$ (equation (12.28)), producing $\rho_j(\rho_1)$, with a standard 4th-order Runge–Kutta integration routine for the differential equation (12.26). Since previous values of f are necessary to include reflections, the obtained values $f(z_j)$ at z_j are stored in an array. By interpolation, these values are used to define $f(z)$ for positive z. We know already that $f(z) = 0$ for negative z.

12.4 Results

We note that the radiated sound field depends on f' and *not* on f itself. So the absolute value of the pressure in the pipe plays a minor role. That is why the wave steepening itself is important, because with increasing (negative) gradients more sound is radiated. Therefore the pipe length is also important to the noise level. In the same way, the explosion time τ (inherent in $\beta(t)$) is important.

For typical values of the parameters ($\beta_1 = 4.5$, an explosion time $\tau = 2.44$ ms, $\Omega = 2092$ cm^3, $L = 60$ cm, $S_j = 9.5$ cm^2, $S_2 = 63.6$ cm^2, $p_\infty = 1$ bar, $\rho_\infty = 1.2$ kg/m^3, $T_\infty = 20°$C, $\gamma = 1.4$, $r = 1$ m) a very representative noise level SPL $= 114.2$ dB is found. The corresponding time histories (time in milliseconds) are given in figure 12.5.

We see in the pot pressure after the initial explosion some residual wiggles of the reflected waves. However small these are, they give rise to some flow through the iris in and out (positive and negative v_2 and v_0). After the initial pulse the decay in the pipe is rather fast, justifying the assumption of weak nonlinearity.

The predicted SPL is exactly in the range of 112–115 dB which is found experimentally. It should be noted that the experiments are, at least under normal operational conditions, not very well repeatable. This is probably because of wind and poorly controllable combustion. So there is evidence that we have indeed modeled the dominating effects. Unfortunately, experimental input of the entropy change β_1 and the explosion time τ is necessary. Therefore, the possibility of studying the effects of pot size and iris diameter is rather limited.

12.5 Conclusions and Suggestions for Further Work

An attempt has been made to model the sound production mechanism of a bird scare gun. The model is not based on a systematic reduction of a more complete model. We have built up our problem description step by step by

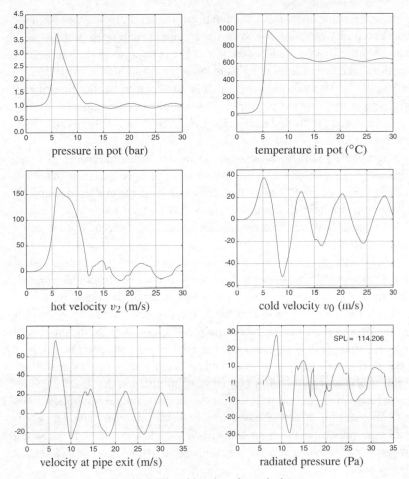

Fig. 12.5. Time histories of a typical case

adding effects and elements like building blocks, until the required accuracy or adequacy is obtained (at least, this was our aim!)

The model consists of four parts. First (1) an increase of energy in the pot is created by combustion of gas. Since combustion is difficult to model from first principles, the resulting increase of entropy is taken from experimental data. Then (2) the high pressure inside the pot drives a hot jet flow from the pot into the pipe. This jet acts like a piston and creates (3) a weakly nonlinear wave in the pipe, which (4) radiates away from the exit to the observer outside.

The results are well in agreement with what could be expected. It is in particular very satisfying to find the resulting SPL to be right among the measured levels. Nevertheless, it is necessarily a little on the high side, because there is no damping at the open-end reflection in our model. In reality the sound wave loses energy by radiation and vortex shedding at each reflection. If we consider only the first few reflections, this effect is not important, but for longer durations, the sound level remains incorrectly at too high a level. It may be noted that the jet from the iris (primarily included to model the generation of the pressure wave) at the same time models a certain amount of vortex shedding at each reflection against the iris. So at the iris there is some loss of acoustic energy.

It is not very difficult to extend the model to include some exit radiation damping on an ad-hoc basis. We simply multiply the terms in formulas (12.13), (12.14), and (12.15) that correspond to the reflected wave by a reflection coefficient R, which is slightly smaller than 1.

In general, however, this reflection involves 3-D effects, which calls for further modelling.

A very important omission in the present model is obviously the combustion. This should be reconsidered seriously, if possible from first principles. Only then is a real study of the effects of iris and pot size possible.

When the pipe length is increased, the predicted noise level increases too, as expected. However, when the pipe is too long, the results become unpredictable. This is probably caused by the wave steepening making the wave more than weakly nonlinear, and probably a fully nonlinear method is necessary. This requires a full numerical solution, though.

An interesting variant, well within the scope of the present model, would be to consider the effects of having a slowly varying pipe diameter (a horn). Until now, the design has been deliberately simple and robust. If the acoustic properties of a variable duct are very favourable, it may become an interesting alternative design.

Acknowledgements

We wish to acknowledge the support from DAZON B.V., Maastricht, The Netherlands, producer (amongst other things) of bird scare guns. We are also grateful for the helpful discussions with dr. Mico Hirschberg (TUE).

References

[1]　Hofmans, G. C. J. (1998) *Vortex Sound in Confined Flows*, PhD Thesis, Eindhoven University of Technology, Eindhoven, Holland.

[2]　Landau, L. D. & Lifshitz, E.M. (1975) *Fluid Mechanics*, Pergamon Press, New York, USA.

[3]   Pierce, A. D. (1981) *Acoustics, an Introduction to its Physical Principles and Applications*, McGraw-Hill, New York, USA.

[4]   Prandtl, L. & Tietjens, O. G. (1957) *Fundamentals of Hydro- and Aeromechanics*, Dover Publications, New York, USA.

Sjoerd W. Rienstra

Department of Mathematics & Computing Science,

Eindhoven University of Technology, P.O. Box 513,

5600 MB Eindhoven, Netherlands

s.w.rienstra@tue.nl

13

Paper Tension Variations

in a Printing Press

Preface

The following case study considers how tension varies in space and time in a continuous sheet of paper (commonly called a *web*) running at high speed through a newspaper press. The model will consider two different parts of the problem in isolation. First, those portions of the paper that are travelling over various rollers are treated, and second, the spans of paper between the rollers are considered. Modelling the first problem requires basic mechanics and considers the forces within the paper and the friction acting against the roller. Practical limiting cases result in a simple ordinary differential equation for the tension as a function of position. When we consider the spans of paper, the model is again mechanical in nature but requires modelling of the stretch within the paper. For practical situations, the resulting model is again an ordinary differential equation which is now time-dependent. The two models are then combined to show how rollers interact and how the tension changes along the paper's length. The mathematics used is relatively simple; the emphasis is on generating a model which is as simple as possible, but which mimics the physical situation. The problem is well suited to mathematical modelling courses because of the wide variety of different aspects of a printing machine that can be considered, and also because the mathematical models that emerge from the modelling process are similar to classical mechanics problems.

The models presented here will require some knowledge of basic mechanics. Hooke's law for a linear elastic material, conservation of mass, and conservation of linear and angular momentum give us the evolution equations of the systems. These approaches lead to linear ordinary differential equations or nonlinear systems of partial differential equations. Nondimensional analysis is a crucial step in simplifying these problems to a manageable form.

The modelling presented here arose from a problem presented by Norm Malmuth of Rockwell International at several Mathematics with Industry

Workshops held at RPI, Troy NY (see [7] and [8]). The problem has been further studied at both the 1992 and 1994 Mathematical Modeling for Instructors courses held at the IMA, University of Minnesota (see [2], [3], and [4]). These reports discuss various aspects of the problem; in addition a more engineering-based approach can be found in [9]. The problem is an excellent teaching tool for getting students to model a physical situation for which some of the resulting mathematics is not too onerous. It has been one of the problems considered as part of an Industrial Case Studies Course within the MSc in Industrial Applied Mathematics at Southampton University. The student's interest can be greatly enhanced by arranging a visit to a local newspaper press, the staff of which are very welcoming to visitors. However, they tend to offer only a tour of the journalistic, editorial, and print-operating process, with the press itself being a minimal part of the visit.

13.1 Problem Definition

In recent years, there has been huge interest in improving the quality of newspapers (for example, through the greater use of colour photographs). Simultaneously, pressure to decrease both production time and cost has mounted. Huge technological strides have reduced the cost of collecting and transmitting the information to be included. In addition, computer technology has transformed the editorial and markup process and produced a situation in which the steps from page layout to the production of the final metal plates for use in a press are completely automated. The plates thus produced are put into high-speed printing presses, which then print, cut, and fold the final newspaper in a single large machine.

To increase printing press throughput, there are a large number of different factors that must be considered. One of these is the tension that is required to drag numerous webs of paper through the press as they pass from printing drums, which imprint the information from the plates, over numerous rollers and cutters into the final folded form. The tension must be both kept sufficiently low to ensure that the paper does not tear but, more importantly, controlled to tight tolerances to ensure that all the printing, cutting, and folding processes stay aligned. Common faults are creases that are formed along the web, misalignment of the multiple passes required to make a colour picture, and misalignment in the cutting of the top and bottom of pages. These problems are believed, at least in part, to be due to changes in the tension of the web. Finding examples of such errors, particularly colour misalignment, in newspapers is relatively easy, and helps motivate students' understanding of the questions of practical importance and the relevance of various factors.

Manufacturers of printing presses constantly consider alternative roller designs that may allow higher paper speeds while keeping the quality adequate. Design engineers would like to have a model of the process which would allow them to consider:

- How will the distribution of tension along a web be changed if an additional conventional static roller is put into the path?
- How can tension changes be modified by changing the surface finish of the roller?
- How will the distribution of tension along a web be changed if an additional free or powered roller is put into the path?
- How do the tension variations propagate through a web of paper when there is a sudden change in the press somewhere? (Such changes normally occur when rolls of paper are changed on the press.)

To answer these questions, we shall first consider the current technology within a printing press and then attempt to derive mathematical models relevant to these questions. In introducing the modelling, there are many aspects of this problem that we shall only touch upon briefly, since in general these require more sophisticated solution techniques – primarily large numerical codes – in order to gain insight into their behaviour.

13.2 Printing Presses

A modern printing press is a very large machine that prints, cuts, collates, and folds an entire newspaper in a single operation. The process is carried out at high speed and with high accuracy. The blank paper is held on huge rolls of paper (called tile rolls). Each is the width of four pages of a standard broadsheet. The machine shown in figure 13.1 is of modular construction and is approximately twenty metres in length and ten metres high. In this configuration it has the capability of feeding four webs of paper through simultaneously to make broadsheet newspapers with thirty-two printed pages, four of which contain full-coloured pictures. The remaining are printed in black.

A significant increase in print speed has been achieved by the automatic changeover and pasting of tile rolls so that empty rolls can be replaced without slowing the printing process. The holders for three rolls for each of the four webs are part of this automatic process. The web is drawn off the roller and passes through the print rollers. For a single-colour imprint, such as black text, a single pass is required. However, a colour print requires four passes (blue, red, yellow, and black) which can be achieved either by use of a single

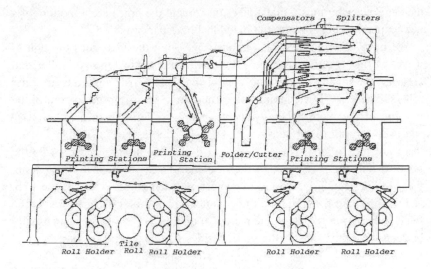

Fig. 13.1. Schematic diagram of a modern printing press

print station with all four drums or by passing the paper through consecutive single or double print stations. When consecutive print stations are used, the web must be kept under carefully controlled tension between the different stations to ensure that the printing is accurately aligned. After printing on both sides of the paper, each web must be cut along its length to make webs of each of the constituent sheets in the newspaper. This cutting of each web is achieved by a splitter. These can be seen in figure 13.1 as points where the web breaks into two. The resulting two webs must then be repositioned so that they run over one another ready for collating. The repositioning of the web is achieved by drag bars (stationary rollers) which are usually placed at an angle of forty-five degrees to the direction of web travel, thus turning the web through right angles. Before collating all the different webs into the final newspaper the individual pages must be aligned. There are two reasons for this. First, all the different pages must be aligned vertically, and second, the entire newspaper must be aligned properly for the cutters that break the web into single newspapers. This alignment process is achieved by use of compensators. These are free-running rollers, usually used in pairs, whose position can be adjusted by the operators. The compensators are seen as a large collection of rollers near the top of the machine. All the webs are now pulled over the "folder". This is a metal surface that puts a single crease into the collection of webs to make the binding of the newspaper. The folded web is then taken by the cutter which chops the web, grasps the bottom of the newspaper by pins (small spikes that create the holes at the bottom of a newspaper), and cuts it into the individual newspapers.

We have now outlined the various stages that the web goes through as it passes through the machine. There are three main types of roller that are used within the machine to control the web. Most numerous are idler rollers. These are free running rollers and are primarily used to control the position of the web. Secondly there are drag rollers which are stationary and used to control the tension in the web, such as at the splitter. Finally there are nip rollers. These rollers are driven by motors and provide the force to pull the web through the machine. The two main nip rollers are the printing drums, which are in contact with the whole width of the web, and the final nip rollers in the folder/cutter which grab only the sides of the folded newspaper (the diamond pattern of their gripping surfaces can usually be seen on either side of the vertical sides of the outer pages of the final newspaper).

The tension within a web can vary as the printing progresses. Control of this is usually achieved by positioning the drag rollers, monitoring the speed of the driving nip rollers, and controlling the tensioners on the large rolls of paper. One of the main sources of tension variation occurs when a tile roll is changed by the automatic mechanism. Although the roll is brought up to the correct speed to minimize the shock through the system, there are large variations as the tensioner adjusts to the different torque required to keep the roll moving.

A visit to a printing press enables students to observe some of the different waves that may be present in the web. Small waves may often be observed running across the web and in some cases creases can be seen on the later parts of the web. Rough observations of the wave size, speed and wavelength can be made. Longitudinal waves cannot be seen and even long transverse waves are almost impossible to observe. The chance to see tile rolls of paper before they are used, to observe the surface texture and finish, and also to see the machine paste a new roll onto the web while in full production allows the student to appreciate the purpose and relevance of the various aspects. A chat with the press operators to discuss what problems they encounter and how they avoid or fix such difficulties is also frequently instructive.

13.3 Modelling

In modeling a web, we consider each of the parts of the process separately. The spans between the rollers are typically of the order of metres in length, and the web is in contact with the roller for around 10 cm. The models will consider the forces acting on the web and the response of the web to these. We will assume that, while the the web is on the roller, it must travel in a circular arc, and from this we will determine how the tension changes around the roller. We will further assume that, while the web is in the span, there are no external forces acting, and hence we can determine the tension and speed in the span. A series of spans will then be considered with each roller providing jump conditions relating to tension and speed.

To gain some insight into the magnitude of the forces acting on the web, it is useful to first consider whether gravitational effects should be included in the model. We need some notation to describe the problem, and accordingly we assume that the roller has radius R; the span between rollers has length L; and the web has density ρ, width w, and (unstressed) thickness h_o. Using physical parameter values (given in table 13.1) associated with a typical paper web, we find that the paper in a one-meter span weighs $\rho w h_o L \approx 1$ kg, which gives a force of 10 N, while typical tensions are $\mathcal{O}(10^2$ N). Hence gravity is not extremely small but can be neglected in an initial model. Over a roller, the gravitational force is an order of magnitude smaller.

13.3.1 Motion over a Roller

We start by considering a model of how the paper passes over the roller. The rollers are metal bars, supported at each end, and long enough for the full width of the paper to touch the bar. The motion of the web would seem to indicate that, as a first approximation, we might assume that the behaviour of the paper is uniform across the width of a roller. This is not entirely supported by observation, since in most cases small waves can be observed running diagonally across the web as well as some disturbances running along the direction of the web. We neglect these nonuniformities to begin with. Thinking ahead, we note that a model requiring transient behaviour in the plane of the web will necessarily require partial differential equations in several space dimensions and hence will be complicated.

The roller is assumed to rotate with angular velocity Ω, giving the surface velocity $U = R\Omega$. We take the web to be very thin so that we can use the independent variable s for the position along the web. We use t for time. As the web moves over the roller, we expect the tension to change and we take the tension in the web $T(s, t)$ as a dependent variable. This tension will cause the web to change its characteristics. Hence we might expect the web to change speed or thin as the tension increases. We will therefore also consider the speed of the web $u(s, t)$, and the thickness of the web $h(s, t)$ as dependent variables.

To create the model, we need to write down conservation laws for mass and momentum, the latter involving any external forces that act on the web.

The basic premise of the roller model is that the velocity of the web is assumed to be purely circumferential, at a radius R, with speed u. Hence we can use cylindrical polar coordinates to write down the relevant equations with $s = R\theta$ and $0 \le s \le L$ being the region where the web is in contact with the roller.

Assuming that the paper has width a, the mass per unit circumferential distance at each point s is $\rho a h$. Hence conservation of mass dictates that

$$\frac{\partial (\rho a h)}{\partial t} + \frac{\partial (\rho a h u)}{\partial s} = 0 . \tag{13.1}$$

It is reasonable to assume that the web material is incompressible and very thin, so that ρ and a can be taken as constant. With this simplification, (13.1) becomes

$$\frac{\partial h}{\partial t} + \frac{\partial (hu)}{\partial s} = 0 . \tag{13.2}$$

The momentum balance equation comes from considering the rate of change of tangential momentum of the web as it passes over the roller. This is balanced by the change in tension on the web and any tangential force (primarily due to friction) between the web and roller. If F is the tangential force, then a balance of the rate of change of momentum gives

$$\frac{\partial (\rho a h u)}{\partial t} + \frac{\partial (\rho a h u^2)}{\partial s} = \frac{\partial T}{\partial s} - F . \tag{13.3}$$

In the usual way, the left-hand side of (13.3) can be simplified using (13.2) to give

$$\rho a h \left(\frac{\partial u}{\partial t} + u \frac{\partial u}{\partial s} \right) = \frac{\partial T}{\partial s} - F . \tag{13.4}$$

In addition we need the momentum equation in the radial direction, and since there is no velocity in this direction (by our assumption), then this gives a simple force balance that the normal force per unit area on the web from the roller is

$$N = \frac{1}{R}(T - \rho a h u^2). \tag{13.5}$$

The first term in (13.5) is due to the tension in the curved web pressing on the roller and the second term is the centrifugal force.

The model equations (13.2), (13.4), and (13.5) are valid for any material moving over a roller. To complete this system, we must supply some model for F to describe how the paper slides on the roller, and for h to determine how the web thins as it extends under tension.

A description of the thinning of the paper comes from experiments and indicates that paper is linearly elastic, that is, it acts like a spring. The stiffness of the spring can be quantified by the Young's modulus, E, of the paper and, because we have assumed the paper to be incompressible, is given by the simple relation

$$h = h_0 - \frac{T}{aE}, \tag{13.6}$$

where both h_0 and E are constants. (In practice these constants would be determined by stretching a length of paper and measuring the length as a function of the applied tension.)

We assume that the dominant tangential force between the roller and the paper is friction. In doing so, we exclude the possibility of lubrication due to either air or moisture trapped between the paper and the roller. Extensive experiments have been carried out to determine this force for different papers over various rollers [5]. Here we will take a simple, but reasonably accurate, description. First we note that friction over a roller acts in the direction opposite to the direction in which the web travels relative to the roller surface so that $F = \kappa(u - U)$ where κ is a positive coefficient. We are taught in basic physics that the friction force between a block and a stationary surface depends only on the normal force N between them. This corresponds to taking

$$\kappa = \mu \frac{N}{|u - U|} \quad \text{so that} \quad F = \mu \frac{u - U}{|u - U|} N, \tag{13.7}$$

where μ is the coefficient of friction (usually taken to be constant). This model is adequate except for the fact that care is needed if $u = U$, since F is then undefined. This difficulty corresponds to defining the static friction. A common model is to say that, while $u = U$, F may take any value, so long as $|F| \le \mu N$. A plot of F as a function of the relative speed $u - U$ is shown in figure 13.2 (broken line). Experimental evidence, such as [5], supports this hypothesis. However, the force is not usually constant when slipping occurs but increases with increasing speed. The vertical part of the graph is therefore usually slightly rounded off (solid line) as a function of velocity.

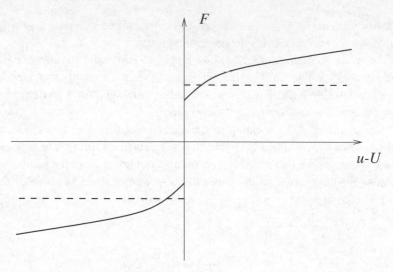

Fig. 13.2. Typical friction force as a function of relative velocity: equation (13.7) (broken line), experimental evidence (solid line)

Equations (13.2), (13.4), (13.7) and (13.6) form a system of hyperbolic equations that can be made dimensionless and then simplified by the introduction of scaled variables. Let

$$h' = \frac{h}{h_0}, \quad u' = \frac{u}{u_0}, \quad U' = \frac{U}{u_0}, \quad T' = \frac{T}{T_0}, \quad t' = \frac{t}{t_0},$$

where the primed quantities are dimensionless and h_0 is the typical thickness of the paper, u_0 the typical speed of the paper, and T_0 the practical tension used. The timescale of interest, t_0, could be taken to be the time for the web to travel around the roller or perhaps the time for the web to travel across a span. Here we take it to be the typical time for the printing press to get from stationary to full running speed. Using these scaled variables, equations (13.2), (13.4) and (13.6) become

$$A\frac{\partial h'}{\partial t'} + \frac{\partial (h'u')}{\partial \theta} = 0,$$

$$Bh'\left(A\frac{\partial u'}{\partial t'} + u'\frac{\partial u'}{\partial \theta}\right) = \frac{\partial T'}{\partial \theta} - \mu\frac{u' - U'}{|u' - U'|}(T' - Bh'(u')^2),$$

$$h' = 1 - KT',$$

where

$$A = \frac{R}{u_0 t_0}, \quad B = \frac{\rho a h_0 u_0^2}{T_0}, \quad \mu, \quad K = \frac{T_0}{E a h_0}.$$

Table 13.1. *Parameter Values*

$h_0 = 10^{-4}$ m	$\rho_0 = 500\,\text{kg}/\text{m}^3$	$u_0 = 30\,\text{m}/\text{s}$
$T_0 = 200$ N	$E = 10^{10} - 10^{12}\,\text{N}/\text{m}^2$ $t_0 = 15$ s	
$a = 2$ m	$R = 10^{-1}$ m $I = 2.5\,\text{kg}\,\text{m}^2$	
$l = 1 - 5$ m		

are dimensionless numbers. Typical parameter values are given in table 13.1, and lead to $A = 1/1500$, $B = 9/20$, $\mu = 0.3 - 0.1$, and $K = 1/100 - 1/10000$. Because later we shall be interested in timescales associated with the dynamics of the span, the roller problem can be considered in a quasi-steady state, where A is taken to be zero. In addition, the stiffness of the web (characterized by the reciprocal of K) is large, so the system can be further simplified by taking $K = 0$. The reduced system is

$$
\left.
\begin{aligned}
\frac{d(hu)}{d\theta} &= 0, \\
Bhu\frac{du}{d\theta} &= \frac{dT}{d\theta} - \mu\frac{u - U}{|u - U|}\left(T - Bhu^2\right), \\
h &= 1,
\end{aligned}
\right\}
\tag{13.8}
$$

where the primes have been dropped for notational convenience. This is the classical "capstan problem" (see for example [6], p.19) whose solution gives the tension change as a rope passes around a ship's capstan. The solution has constant velocity u and constant web thickness h with the tension (for $u > U$) given by

$$
T = Bhu^2 + be^{\mu\theta}.
$$

The tension change around the roller is therefore exponentially related to the angle of contact between the web and the roller. For the purposes of modelling the web motion as a whole, the behaviour of the web around the roller can be summarized as jump conditions between the point where the web first reaches the roller (indicated by $^-$) and the point where it leaves it (indicated by $^+$). We have

$$
u^+ = u^- , \qquad h^+ = h^- ,
$$

$$
T^+ - Bh^+(u^+)^2 =
\begin{cases}
\left(T^- - Bh^-(u^-)^2\right)e^{-\mu L/R} & \text{for} \quad u < U, \\
\left(T^- - Bh^-(u^-)^2\right) & \text{for} \quad u = U, \\
\left(T^- - Bh^-(u^-)^2\right)e^{\mu L/R} & \text{for} \quad u > U.
\end{cases}
$$

$$
\tag{13.9}
$$

The model of the tension variations around the roller is therefore complete. There are some difficulties with considering variations in the web motion around the steady motion $u = U$ due to the discontinuity in (13.9). These can be avoided either by considering a smooth friction curve or allowing K to be small but nonzero. The $K \neq 0$ analysis is the same as above if the web is either faster than or slower than the roller at all points. However, for $u \approx U$, two regions of the roller must be considered. In one region, $u = U$ and there is no frictional force; in the other, $u \neq U$ and a frictional force is present (corresponding to a partially slipping web). The details of the analysis are left as an exercise. For the present case, the final jump conditions may be summarized as:

$$h^+ u^+ = h^- u^-, \qquad h^+ = 1 - kT^+, \qquad h^- = 1 - kT^-,$$

$$T^+ - Bh^+(u^+)^2 = \begin{cases} \left(T^- - Bh^-(u^-)^2\right) e^{-\mu\phi} & \text{for} \quad u^+ < U, \\ \left(T^- - Bh^-(u^-)^2\right) & \text{for} \quad u^+ = U, \\ \left(T^- - Bh^-(u^-)^2\right) e^{\mu\phi} & \text{for} \quad u^+ > U. \end{cases}$$

$$(13.10)$$

Here ϕ (the slip angle) corresponds to the angle around the roller where the web first slips relative to the roller, and is chosen by considering the two cases $\phi = R/L$ (corresponding to full slip) and $u^- = U$ (corresponding to partial slip) and then selecting the case which satisfies $(u^- - U)(u^+ - U) \geq 0$ (corresponding to friction acting against the relative motion of the web to the roller).

The modelling of the roller is completed by considering the dynamic motion of the roller due to the paper tension acting upon it. Within the printing press there are numerous drag rollers, where the roller is held stationary. There $U = 0$. Also, there are nip rollers, where U is specified by the driving motor speed. However, there are far more idler rollers that are free to rotate and accelerate to near the paper speed. A simple model of such a free roller may be proposed by balancing the rotational inertia of the roller (they are typically made of steel) with the tension change in the paper each side of the roller and any losses due to viscous effects in the bearings holding the roller. A force balance then dictates that the surface speed U of the roller is governed by

$$\frac{I}{R}\frac{dU}{dt} = \int_0^L F\, ds - \frac{\nu U}{R},$$

where I is the moment of inertia of the roller, $\nu U/R$ is the loss rate in the bearing, and F is the friction force due to the slipping paper. Using the previous analysis to determine the friction force, the force balance (in nondimensional units) becomes

$$\beta \frac{dU}{dt} = T^+ - Bh^+ (u^+)^2 - T^- + Bh^- (u^-)^2 - \alpha U, \tag{13.11}$$

where the nondimensional parameters are

$$\beta = \frac{Iu_0}{R^2 T_0 t_0} \quad \text{and} \quad \alpha = \frac{\nu u_0}{R^2 T_0}.$$

13.3.2 Motion in a Span

Between rollers, we can assume that the web has no external forces acting upon it. (This is reasonable save for the fact that the air flow induced by the web motion can create forces that may create waves.) In dimensionless form, the mass conservation equation for the web is given by

$$\frac{\partial h}{\partial t} + \frac{\partial (uh)}{\partial x} = 0, \tag{13.12}$$

where x measures distance along the path of the web between rollers (which we assume is a straight line), and the momentum balance is

$$\rho a h \left(\frac{\partial u}{\partial t} + u \frac{\partial u}{\partial x} \right) = \frac{\partial T}{\partial x}. \tag{13.13}$$

In addition the equation of state of the web (13.6) still holds. If this is used to eliminate T, then the resulting system (13.12) and (13.13) describes the elastic waves that propagate up and down the web. The characteristics of this system are familiar from the theory of shallow-water waves (see, for example [1] pp. 89–92) and are associated with a speed $u \pm \sqrt{E/\rho}$. Using the typical values given in table 13.1 it follows these waves travel at around 5000 m/s which is many times faster than the web speed. Hence these elastic waves will traverse the span many times before the web crosses the span and hence the tension will have plenty of time to equilibrate. Thus in each span the tension is uniform in space but may vary in time. We denote the tension in the nth span by $T_n(t)$. Finally, in each span as the tension changes the total mass of web in the span will change due to changes in the web thickness. Integrating (13.12) along the span gives

$$L \frac{dh}{dt} + uh \Big|_{x=0}^{x=L} = 0.$$

This equation, together with the assumption that the tension is a function of time only, and the constitutive equation of the web (13.6) give the conditions in each span. The previous informal arguments can be put into a more formal structure in terms of asymptotic analysis, but this is left to the interested reader.

13.4 The N-Roller Start-up Problem

In a typical configuration, the web passes over a large number of rollers as it passes through the printing press, and it is of interest to determine the overall dynamics of this web–roller system. The previous sections derived the equations that describe each roller and each span. Consider the rollers to be numbered $1, 2, \ldots, N$, and the span between rollers n and $n + 1$ to be numbered n. Rollers may have different moments of inertia and bearing losses, and spans may be of different lengths. Hence the equations to be solved, in nondimensional form, are

$$\gamma_n \frac{dh_n}{dt} + h_n \left(u_{n+1}^- - u_n^+\right) = 0, \qquad n = 1, 2, \ldots, N - 1,$$

with the equations for the roller velocities

$$\beta_n \frac{dU_n}{dt} = \left(T_n - Bh_n(u_n^+)^2\right) - \left(T_{n-1} - Bh_{n-1}(u_n^-)^2\right) - \alpha_n U_n,$$

$$n = 1, 2, \ldots, N,$$

the jump conditions

$$h_n u_n^+ = h_{n-1} u_n^-,$$

$$T_n - Bh_n(u_n^+)^2 = \left(T_{n-1} - Bh_{n-1}(u_n^-)^2\right) \exp\left(\mu \phi_n \, \text{sgn}(u_n^- - U_n)\right),$$

and the equation of state

$$h_n = 1 - KT_n .$$

In these equations, to simplify notation, the parameters $\gamma_n = L_n/(u_0 t_0)$ describe the nondimensional span distances, and the function $\text{sgn}(z)$ is defined by

$$\text{sgn}(z) = \begin{cases} 1 & (z > 0), \\ 0 & (z = 0), \\ -1 & (z < 0). \end{cases}$$

Initial conditions must be prescribed indicating the tension in the spans and the speed of the rollers (perhaps by specifying h_n and U_n). In addition, boundary conditions are needed describing how the web is attached as it enters and leaves the printing press. On the feed into the press, sensors are installed to help ensure that the paper is pulled off the rolls at constant tension. Hence T_0 should be given. At the last part of the processing of the paper in the printing press, the web is pulled through by nip rollers travelling at a prescribed speed. Hence u^+ should be specified. This gives a complete specification of the problem.

The evolution equations for h_n and U_n (the components of the vectors h and U) can be solved numerically using a combined implicit–explicit time integration scheme. The form of the equations is

$$\gamma_n \frac{dh_n}{dt} = F_n(\mathbf{h}, \mathbf{U}), \qquad \beta_n \frac{dV_n}{dt} = G_n(\mathbf{h}, \mathbf{U}).$$

This assumes that all the values u_n^- and u_n^+ can be found in terms of $\mathbf{h}$ and $\mathbf{U}$ from the jump conditions and the equation of state. Because γ_n is typically much less than β_n, h_n is a "fast" variable compared to U_n. Thus, it is best to handle the equation for h_n implicitly, while the equation for U_n can be handled explicitly as before. For example, a simple first-order scheme is

$$\gamma_n \frac{h_n^{(k+1)} - h_n^{(k)}}{\Delta t} = F_n(\mathbf{h}^{(k+1)}, \mathbf{U}^{(k+1)}),$$

$$\beta_n \frac{U_n^{(k+1)} - U_n^{(k)}}{\Delta t} = G_n(\mathbf{h}^{(k)}, \mathbf{U}^{(k)}),$$

where the superscript (k) denotes the discrete variables at a time $t = k\Delta t$. This scheme was used in the following numerical examples.

First, the problem of three identical rollers with two equal-length spans was considered. Physically relevant parameters were used, which resulted in the nondimensional parameters taking the values $N = 3$, $\alpha = 0.1$, $\beta = 2.5$, $\mu = 0.3$, $K = 0.01$, $L/R = \pi/3 \sim 1.0472$, and $B = 0.45$. The boundary conditions were taken to be $T_0 = 1$ and $u_3^+ = 1$. This was run to steady state and resulted in the values given in table 13.2.

Table 13.2. *Steady-State Values*

n	V_n	ϕ_n	h_n	v_n	T_n
0			0.9900	0.9970	1.0000
1	0.9970	0.5464	0.9890	0.9980	1.0997
2	0.9980	0.5222	0.9880	0.9990	1.1995
3	0.9990	0.4121	0.9870	1.0000	1.2994

We note that the losses due to the bearings are very small and that the speeds of the roller surfaces are very close to the web speed. As a result, over each roller, approximately half of the surface has the web slipping and half has it not slipping.

Ideally, the web should run at steady-state conditions. However, perturbations inevitably arise, and it is of interest to determine the response of the web and rollers to these perturbations. One way to model disturbances to the steady-state web is to perturb the boundary conditions. The web tension at the inlet and/or the web velocity at the outlet may be changed. Let us consider perturbations $f(t)$ from the steady-state constant value at the boundary of the general form

$$f(t) = \begin{cases} 1 - \epsilon \sin^2\left(\frac{\pi t}{t_1}\right) & (0 \le t \le t_1), \\ 1 & (t > t_1). \end{cases}$$

In this form, the disturbance has a finite duration and an amplitude ϵ. This general form is applied to either the inlet tension T_0 or the output velocity u_N^+ for $t \ge 0$.

A system of three rollers with four spans is considered from steady-state, as given in table 13.2 using the data $\gamma_1 = \gamma_2 = 0.00667$ and $t_1 = 0.1$. A numerical time step of $\Delta t = 0.0005$ was used for both calculations. Figure 13.3 shows the response of the rollers to a small perturbation in inlet tension ($\epsilon = 0.2$), and figure 13.4 shows the response of the span tensions to the same perturbation. In figure 13.5 the inlet tension perturbation is much larger ($\epsilon = 0.4$). For all cases, the disturbance at the inlet propagates down the web from roller 1 to roller 2 to roller 3 with a decrease in amplitude, and the response then decays to the steady state. For the case $\epsilon = 0.4$, the perturbation has sufficiently large amplitude to cause rollers 1 and 2 to change from partial to full slip and then back to partial slip as they return to steady state. This behaviour is best seen in the plot showing the web slip angles in figure 13.5.

Figure 13.6 show the response of the rollers to velocity perturbations at the outlet from a steady state with $\epsilon = 0.004$. From these plots, it can be seen that the web–roller system is much more sensitive to perturbations in the prescribed web velocity at the outlet than to tension changes at the inlet.

As a final note, the response of a system with ten rollers to an input tension perturbation is shown in figure 13.7. For this calculation, all conditions are the same as earlier except $N = 10$. In this figure, a clear propagation speed is seen as the disturbance travels through the roller–web system. Such a propagating

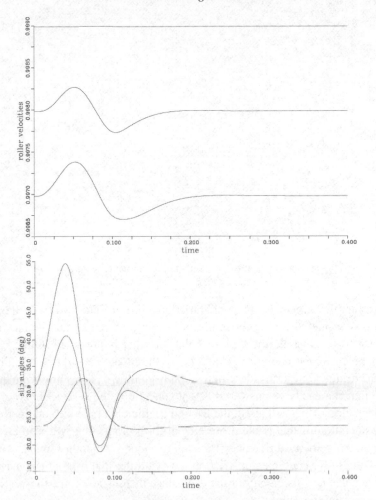

Fig. 13.3. Three roller velocities and slip angles during velocity perturbation of the input from steady state with, $\epsilon = 0.2$, $\gamma_1 = \gamma_2 = 0.00667$, and $t_1 = 0.1$

wave could of course be studied by considering a large number of rollers and analyzing the resulting continuum model.

13.5 Concluding Remarks

This case study has presented a simple model of dynamic web motion over a number of rollers. The model has been broken into two parts. The first part considers the web as it passes over each roller; here the motion is quasi-static and leads to jump conditions for the tension and velocity across such a roller. The second part concerns the motion in a span; here the motion is

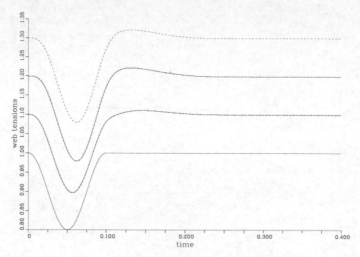

Fig. 13.4. Four span tensions during velocity perturbation of the input from steady state, with $\epsilon = 0.2$, $\gamma_1 = \gamma_2 = 0.00667$, and $t_1 = 0.1$

nearly uniform in space but with variations in time. The two basic models can be considered for various collections of rollers. One example has been presented to illustrate the type of behaviour that is predicted. Other cases may be examined numerically, and hence the model can address a number of the original questions identified at the beginning of this study. One question that can easily be answered is that concerning the surface finish of the rollers; clearly this will affect the friction coefficient of the web on the roller. Consideration of this is therefore crucial for a drag roller, where there are significant velocity differences. For idler rollers, a smoother finish, with a resulting lower coefficient of friction, gives a roller that is less responsive to sudden changes in velocity of the web, since there is less force to overcome the inertia of the roller.

Within the text, several indications have been given as to how the model can be extended. Additional modelling and analysis studies that may help to understand the behaviour of the printing press might include the following features.

- Inclusion in the model of transverse waves as well as the longitudinal waves that have already been considered. Such transverse waves propagate much more slowly than longitudinal waves, and their speed may become comparable to the web speed. In these circumstances, we may expect many complicated effects such as wave reinforcement to occur. These effects may be particularly important in the region where the web approaches a roller.

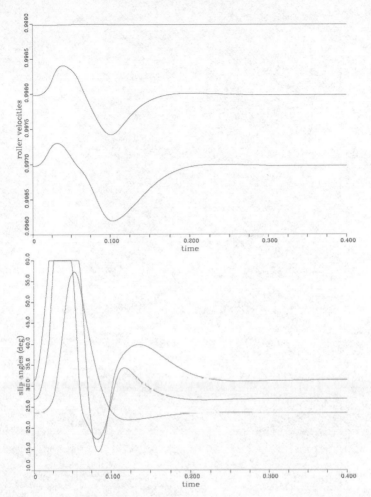

Fig. 13.5. Three roller velocities and slip angles during velocity perturbation of the input from steady state, with $\epsilon = 0.4$, $\gamma_1 = \gamma_2 = 0.00667$, and $t_1 = 0.1$

- Performing a linear stability analysis to verify the numerical calculations.

- Carrying out additional analysis to determine analytically the propagation speed for the disturbances as they travel through the system.

- Considering a model of a control system so that the interaction between the web and its control system can be considered.

- Taking into account the fact that, in practice, numerous webs run simultaneously through the printing press. These webs move separately until they reach a position near to the folder, whereupon they run over each

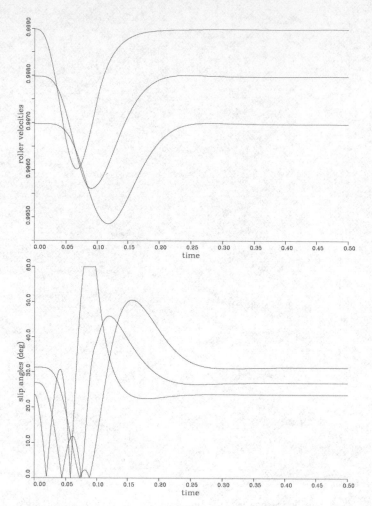

Fig. 13.6. Three roller velocities and slip angles during tension perturbation of the output from steady state, with $\epsilon = 0.001$, $\gamma_1 = \gamma_2 = 0.00667$, and $t_1 = 0.1$

other and over rollers. The friction between the various webs and the friction with the roller needs to be included in the model and the effects on tension variations considered.

Acknowledgements

The work presented here is the product of the efforts of numerous researchers who attended the RPI and IMA meetings. Thanks are due to Norm Malmuth,

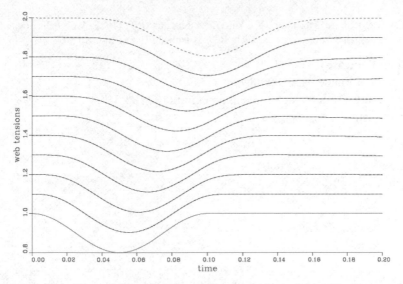

Fig. 13.7. Ten roller velocities during tension perturbations of the input from steady state, with $\epsilon = 0.2$, $\gamma_1 = \gamma_2 = 0.00667$, and $t_1 = 0.1$

Julian Cole, Pat Hagan, John King, and in particular to Don Schwendeman for assistance.

References

[1] Acheson, D. J. (1990) *Elementary Fluid Dynamics*, Clarendon Press, Oxford, England.

[2] Cole, J., Adkins, F., Deng, Z., Ji, C. & Mukherjee, B. (1992) A wrinkling criteria equation for web spans. Mathematical Modeling for Instructors, IMA Preprint Series 1021, Univ. Minn. USA.

[3] Cole, J., Buchanan, J. R., Cao, G. & Lai, G. (1992) A model of tension and velocity in web handling machinery. Mathematical Modeling for Instructors, IMA Preprint Series 1021, Univ. Minn. USA.

[4] Gupta, C. P., Kuller, R. G., Ma, X., McClure, P., Morlet, A. C., Please, C. P., Salmassi, M., Struther, A. A. & Valdivia, R. (1994) Zen and the art of printing press control, Mathematical Modeling for Instructors, IMA Preprint Series 1254, Univ. Minn. USA.

[5] Kornmann, P. (1975) Determination of the coefficient of friction of web-guiding cylinders during continuous transport of printing materials. Papier und Druck **27**, 42–46.

[6] Kreyszig, E. (1993) *Advanced Engineering Mathematics, 7th. Edn.*, John Wiley & Sons, New York, USA.

[7] Schwendeman, D.A. (Ed.) (1993) Mathematics with Industry Workshop Report, RPI, Troy, NY USA.

[8] Schwendeman, D.A. (Ed.) (1994) Mathematics with Industry Workshop Report, RPI, Troy, NY USA.

[9] Taguchi, T., Morimoto, K., Yuji, T. & Sakamoto, Y. (1985) Web behaviour on newspaper rotary presses. Tech. Rev. Mitsubishi Heavy Ind. **22**, 303–308.

Colin P. Please
Faculty of Mathematical Studies
University of Southampton, Southampton SO17 1BJ, UK
cpp@maths.soton.ac.uk

Index

P_1-approximation, 181
$\mathcal{L}_{[a,b]}$, 246
4th-order Runge–Kutta method, 266

absorbed radiation, 182
absorption coefficient, 183, 191
accelerometer, 237
acid, amino, 199
acoustic approximation, 259
acoustic emission, 80
acoustic emission testing, 81
acoustic energy, 91
acoustic energy sink, 258
acoustic perturbation, nonlinear, 263
activation energy, 143, 164
activation energy, large, 155
activation temperature, 164
active site, 200
actuators, piezoceramic, 229
adiabatic, 33
advection, 118
advection coefficient, 128
advective flux, 163
Aeronautical Research Council, 1
agent, precipitating, 200
air compression, 24
air damping coefficient, 234
air lubrication, 277
air-conditioners, 24
airfields, 253
algebraic equation, 147, 187, 265
Allied Signal, 135
alloys, shape-memory, 229
aluminum alloy tank, 90
amino acid, 199
amplitude equations, 218
analysis, functional, 245
angle, contact, 205
angle, solid, 183, 187
animal dynamics, 115
annealing process, 182
antibodies, maternal, 130
antigen immunity, 130

ANZIAM, 2
apparent gas velocity, 139
approximation, acoustic, 259
approximation, first-order, 190
approximation, Galerkin, 247
aqueous solution, 199
Arrhenius law, 104, 163, 176
ASME pressure vessel code, 81
asymptotic expansion, 191, 218
asymptotic limit, 166
asymptotic techniques, 137
attenuation, 183
autofrettage, 81

backplate, 23
baked electrode, 47
band, spectral, 186
beam theory, 66
bearing loss rate, 281
bearings, rubber, 229
belts, 229
Bernoulli's equation, 14, 74, 256, 258
Bernoulli–Euler bending theory of thin plates,
    72
BFGS optimization routine, 243
biharmonic equation, 18
binary diffusion coefficient, 204
binder, 48
binoculars, astronomical, 181
biochemistry, 199
biocide, 117, 132
biological control, 116, 117
biology, structural, 199
biomolecules, 199
biorthogonal functions, 59
bird scare gun, 253, 266
black body boundary condition, 196
black body intensity, 184, 186
blast furnace, 160
blow up, finite-time, 146
blowflies, 117
Boltzmann constant, 183
Boltzmann integral model, 240

boundary condition, black body, 196
boundary condition, clamped, 237
boundary condition, diffuse, 196
boundary condition, Neumann, 163
boundary condition, quasi-stationary, 222
boundary condition, specular, 196
boundary layer, 17, 58, 143, 168, 191, 210
boundary layer, inlet, 169
boundary layer, thermal, 162
boundary layer, viscous, 162
boundary radiative flux, 186
brain storming, 2
breakfast cereal, 97
breathing air cylinders, 80
breeding rate, 119
bridges, 229
British Steel Corporation, 160
building foundations, 229
buoyancy, 210
butane, 253, 255

calcinated anthracite, 48
calcium carbide, 46
capstan problem, 279
carbon black, 229
carbon dioxide, 135
carbon electrode, 47
carbon monoxide, 135, 141
carbon paste, 46
carrots, chopped, 117
catalyst pellets, 161
catalytic converter, 135
Cauchy–Green tensor, 232
causality, condition of, 260, 264
cellular immunity, 130
centrifugal force, 277
ceramic, 196
ceramic monolith, 135
cereals, 97
chemical bond, rotation of, 200
chemical energy, 254, 258
chemical equilibrium, 200
chemical reactions, 135, 137
child's swing, 70
choke volume, 24, 36
circle involutes, 22
clamped boundary condition, 76, 237
Claremont Colleges, 2
Claremont Mathematics Clinic, 66
classical mechanics, 270
coated grains, 98
coefficient of friction, 277
coefficient of thermal expansion, 69
coefficient, air damping, 234
coefficient, heat transfer, 162, 166
coefficient, Kelvin–Voigt, 248
coefficient, scattering, 197

coke, 48
cold branch, 148
collocation, 59
color misalignment, 271
color photographs, 271, 272
combustion, 254, 267
combustion chamber, 253
commercial software packages, 181
compensators, 273
components, elastomeric, 229
compressed gas, 24
compressible lubrication theory, 22
compressible Navier–Stokes equations, 26
compressible turbulent mixing, 254
compression ratio, 25, 38
computer graphics, 187
condensation, 225
conductive heat transfer, 163
conductivity, thermal, 166, 183, 204
configuration map, 232
conjugate-gradient method, 219
conservation law, differential, 5, 139
conservation of angular momentum, 270
conservation of linear momentum, 270, 276
conservation of mass, 26, 139, 270, 276
constitutive equation (of newspaper web), 282
constitutive law, hysteretic, 237, 240
contact angle, 205
continuous carbon electrode, 47
continuous nonlinear mapping, 247
convection, 138
convective derivative, 26, 51
convergence, exponential, 219
corn grains, 97
corn kernel, 104
corner nodes, 60
corner solutions, 54
Couette flow, 29
crack length, 90
creases (in newspaper web), 271
critical points, 92
CRSC (Center for Research in Scientific
    Computation), 229, 248
crystal growth, protein, 199
crystal nucleation, 227
crystalline phase, 199
CSIRO (Commonwealth Scientific and
    Industrial Research Organisation), 116
cyclical tests, 236

d'Alembert's solution of the wave equation,
    260
damping, Kelvin–Voigt, 234, 245, 248
DANFOSS, 22
DAZON B.V., 268
dead zone, 171
decibels, 254

deformation gradient, 232
Deissler jump condition, 190, 191
delta function, 118
densitometer, 66
depletion of reactant, 163
derivative, Frechet, 247
design of drugs, 199
detonation, 253
diagonal web waves, 275
differential equation, 135, 147, 151, 192, 193, 202, 265, 270
diffraction, x-ray, 200
diffuse boundary condition, 196
diffusion, 5, 118
diffusion approximation, 189, 196
diffusion coefficient, 103, 128
diffusion coefficient, binary, 204
diffusion of vapor, 199, 200
diffusion-limited heat transfer coefficient, 222
diffusive transport, 205
dilatational viscosity, 26
discrete ordinates, 181, 189
discretization, 187, 188
disease, 115, 200
dispersion constant, 115
dissipation, 26
distributional sense, 242
double pendulum, 70
double print station, 273
drag bars, 273
drag rollers, 274
dried grains, 98
drug design, 199
duality product, 246
dynamic behaviour, 137
dynamic experiments, 235
dynamic web motion, 285
dynamical system, 35

ECMI, 1
editor, 271
editorial mark-up, 271
effective thermal conductivity, 162
eigenvalue, 149
eigenvalue problem, 52, 53, 149, 152, 156
elastic plate, 66
elastic waves, 281
elastomers, 229, 230, 237
electric smelting, 46
electrode consumption, 47
electromagnetic spectrum, 182
element, finite volume, 187, 189
elements, macro, 219
ELKEM ASA, 46
elongation, simple, 233
emissivity, 163
emitted radiation, 182

end correction, 260
endosperm, 99
endothermic reaction, 160
energy equation, 26
energy flow, radiative, 183
energy integral, 258
energy-preserving discretization, 187
energy, activation, 164
energy, chemical, 254, 258
energy, kinetic, 33
energy, thermal, 27, 33, 256
engine exhaust, 135
enthalpy, 33, 207
entropy, 36, 256
enzymes, 97
EPSRC, 4
equation of state, 27
equilibration of tension, 281
equilibrium points, 149
equilibrium, chemical, 200
equilibrium, thermal, 162
ESGI (European Study Groups with Industry), 160
Euler equations, 13
Euler method, 188
Eulerian formulation, 232
European Study Group with Industry, 46
evaporation, 205, 220, 223, 224, 227
existence, solution, 245
exothermic reaction, 137, 160
expansion coefficient, volumetric, 204
experimental matrix, 226
experiments, 3
experiments, dynamic, 235
explicit-finite difference method, 124
explicit Lagrangian equations, 61
exponential convergence, 219
exponential decay of intensity, 191
exponential evaporation curve, 220
exponential memory kernel, 242
exponentially small reaction rate, 143
extension, simple, 230

faecal/oral spread, 117
fast evaporation limit, 223
fast time scale, 211
Fastflo, 105
fenders, 229
ferro-alloys, 46
fibre reinforced pressure vessels, 80
Fick's law, 7
finite-difference method, explicit, 124
finite-difference method, implicit, 124
finite differences, 160
finite-element method, 60, 202, 219
finite memory, 241
finite-strain elastic theory, 231

finite-strain theory, 230
finite volume method, 187, 189
finite-time blow up, 146
fire-fighters, 80
first moment, 190
first-order approximation, 190
first-order expansion, 192
first-order ODE, 152
first-order reaction kinetics, 161
first-order scheme, 283
fixed base evaporation, 205
flaked grains, 98
flame front propagation, 254, 256
fluid compressibility, 69
flux, advective, 163
flux, of momentum, 259
flux, radiative, 183, 184, 194
Fokker–Planck derivation of diffusion
    equation, 119
foot and mouth disease, 118
force balance, 276
force, centrifugal, 277
FORTRAN, 243
Fourier transform, 123
Fourier's law, 183
Fox, Leslie, 1
fractures, 181
Frank–Kamenetskii approximation, 164
Fraunhofer Institute, 181
Frechet derivative, 247
free oscillations, 74
free-release experiment, 237
friction, 270, 276
friction coefficient, 277
friction, static, 277
Froude number, 50
FRP, 80
FS (finite-strain) theory, 230
functional analysis, 245
functional, nonlinear, 247

Galerkin method, 235, 247
gas constant, 27, 164, 256
gas explosion, 253
gas law, ideal, 27, 256
gas stream, 161
gas viscosity, 162
gas, perfect, 27, 257
gas–solid reaction, 160
gasoline, 66
Gauss formula, 187
Gauss numerical quadrature, 219
gelatinisation, 97, 100
Gelfand triple, 246
generic oxidand, 141
genetic sequence, 199
geometry of plane curves, 25

glass, 181
glass cylinder, heated, 195
glass industry, 190
glass production, 181
global weak solution, 247
grades, student, 2
gradient, deformation, 232
grain composition, 98
grain cooking, 97
grain swelling, 113
grain wetting, 97
Green–St.Venant strain, 232
group, non-dimensional, 213, 227
guitar string, 70

heat absorption, 176
heat and mass transfer, 138
heat capacity, 256
heat conduction parameter, macroscopic, 162
heat diffusion, 27, 97, 136
heat flux vector, 183
heat flux, radiative, 175
heat of reaction, 162
heat transfer coefficient, 139, 162, 166, 206,
    222
heat transfer equation, 182
heat transfer problem, radiative, 160
heat transfer, conductive, 163
heated glass cylinder, 195
heating oil, 66
Heaviside function, 156, 177
highly filled rubbers, 249
Hilbert space, 246
homentropic flow, 27
Hooke's law, 19, 270
hoop stress, 84
horn, 268
hot branch, 148
hot–cold mixing zone, 255
hydrocarbons, 136
hydrogen vehicles, 80
hyperbolic equations, system, 278
hysteresis, 229, 237, 240
hysteresis loop, 241
hysteretic constitutive law, 237, 240

ideal gas law, 27, 256
idler rollers, 274
ignition temperature, 143, 156
IMA, University of Minnesota, 271
implicit finite-difference method, 124
implicit-explicit numerical scheme, 283
impulse response experiment, 237
impulsive noise, 253
index, refractive, of glass, 183, 195
index, spectral radiative, 183
Industrial Case Studies Course, 271

industrial refrigerators, 24
inertia, rotational, 280
infected death rate, 115
infection diffusion, 118
infective, 118
infinite-dimensional state space, 235
inlet boundary layer, 169
inner solution, 147
Instron machine, 235, 241, 243
integrals, numerical evaluation of, 150
intensity, black body, 184
intensity, exponential decay, 191
intensity, radiative, 185, 189
interface heat transfer coefficient, 206
interface vapor pressure, 226
interface velocity, 205, 225
intermediate time scale, 211
internal stress field, 182
interpolation, 188, 266
invariants, strain, 231
inverse problem, 236, 237, 243
iris, 253
isentropic flow, 256
isotherm, Langmuir, 176
isothermal, 32, 43
isotropic scattering, 197
iterative conjugate gradient method, 219
ITT-Barton, 66
ITWM, 181

jet, quasi stationary, 256
jet, turbulent, 258
journalist, 271
jump condition, Deissler, 190, 191
jump conditions, 275, 279
jump term, 247

Kelvin–Voigt damping, 234, 245, 248
kernel, memory, 241
kerosene, 66
kill rate, 117
kinematic condition, 60
kinetic evaporation path, 227
kinetic path, 202
kinetics, of supersaturation, 227

Lagrangian formulation, 60, 232
laminar flow, 162
Langmuir isotherm, 176
Laplace's equation, 71, 73
leakage of gas, 24
leakage rate, 22
length scales, multiple, 191
lenses, 181
Lewis number, 213
light speed, 183
light-off, 144, 146

light-off point, 136, 143
lightweight pressure vessels, 80
Linbro plate, 226
linear elastic shell theory, 85
linear elasticity, 95, 270
linear splines, 235, 236, 248
load cell data, 248
local coordinates, 51
local thermal equilibrium, 176
London Mathematical Society, 4
Lord Corporation, 229, 248
loss rate, in bearing, 281
low cycle fatigue, 81
lubrication theory, 28, 30
lumped parameter model, 71

Mach number, 255
macro elements, 219
macromolecule, protein, 210
macroscopic heat conduction parameter, 162
MAF (Ministry of Agriculture and Forestry), 116
magnetostrictive core, 67
map, configuration, 232
MAPLE, 80, 94
mass conservation, 26, 33, 270, 276
mass transfer coefficient, 139
matching, 149, 168, 174, 175
maternal antibodies, 130
Math Clinics, 1
Mathematica, 223
mathematical biology, 115
Mathematical Modeling for Instructors course, 271
MATLAB, 243, 248
matrix, experimental, 226
Maxwell, J.C., 224
mean action time, 101
mean free path, 182
mechanics, classical, 270
memory kernel, 241, 242
memory, finite, 241
metal press plates, 271
method of multiple scales, 199, 203
MFLOPS, 202
midpoint rule, 151
midside nodes, 60
Minty–Browder technique, 247
MISG (Mathematics-in-Industry Study Group), 97
mixing zone, hot–cold, 255, 259
mixing, turbulent compressible, 254
mixture, stoichiometric, 253
modeling, pseudo–phenomenological, 241
Moffatt vortex, 54
momentum conservation, 276
momentum flux, 259

monotone decreasing trajectory, 174
monotonicity condition, 247
Mooney SEF model, 231
MPI (Mathematical Problems in Industry Meeting), 135, 271
MSc in Industrial Applied Mathematics, 271
multi-dimensional parameter space, 200
multi-processor machine, 219
multiple length scales, 191
multiple scales method, 199, 213, 214, 223
myxomatosis virus, 116

NACA/NASA, 1
NaCl, 200
NAG library, 106
natural frequency, 70, 72, 74
natural gas, 80
natural gas vehicles, 80
Navier–Stokes equations, 12
Navier–Stokes equations, 26, 50, 203
necking, 57
Nekton, 202, 219
Nelder–Mead optimization routine, 243
Neo-Hookean constitutive law, 248
neo-Hookean materials, 231
Neumann boundary condition, 31, 163
New Zealand pastures, 116
newspaper press, 270, 271
Newton boundary condition, 206
Newton's method, 154
nip rollers, 274
nitrous oxide, 137
NMR moisture profiles, 113
no slip condition, 50
noise level, 268
noise, impulsive, 253
nominal stress, 233
noncanonical end problem, 59
nondimensionalisation, 30, 50, 213, 227, 270
nonhomogeneous thermal contraction, 181
nonuniform asymptotic expansion, 218
nondestructive testing, 80
nonlinear acoustic perturbation, 263
nonlinear diffusion, 97, 105
nonlinear dynamics, 115
nonlinear functional, 247
nonlinear mapping, continuous, 247
nonlinear partial differential equation, 234, 235, 270
nonlinear wave steepening, 253
nonmonotonic behaviour, 80, 90
North Carolina State University, 230
nucleation (of crystal), 227
numerical analysis, 46
numerical evaluation of integrals, 150
numerical methods for PDE's, 187
numerical scheme, implicit-explicit, 283

numerical simulation, 37, 136, 199
Nusselt number, 206

oil industry, 66
oil-fields, 253
one scale asymptotic, 191
one-dimensional wave equation, 260
opaque, 182, 185
operator, variational projection, 219
optimization routine, BFGS, 243
optimization routine, Nelder-Mead, 243
orchards, 253
ordinary differential equation, 22, 34, 151, 152, 187, 223, 270
ordinates, discrete, 181, 189
outer solution, 147, 168
outgoing wave, 261
overcooking, 97, 109
overwrapped cylinders, 80
Oxford University, 1
oxidands, 138, 141
oxidation reaction, 135, 139

paper, 270
Papkovitch–Fadle equation, 59
parameter estimation techniques, 243
parameter identification problem, 248
parameter space, 200, 236
partial differential equations, 72, 130, 160, 187, 202, 234, 235, 245, 275
particle Reynolds number, 162
paste segregation, 48
path, kinetic, 202
Peclet number, 162
Peclet number, reduced, 29
pellets, catalyst, 161
peptide bond, 199
perfect gas, 27, 257
pericarp, 99
periodic pressurization, 81
peristaltic action, 27
perturbation analysis, 189
pH, 201
phase, crystalline, 199
photographs, color, 271
photon, 182
piecewise linear splines, 237
piezoelectric crystal, 67
piezoceramic actuators, 229
pigs, feral, 118
PIMS, 1, 80
pins, 273
Piola–Kirchhoff stress tensor, 233
piston, 253
pitch, 48
Planck's function, 183
plastic deformation, 80

plasticity test, 48
platinum, 135
point source spread, 118
Poiseuille flow, 29
poisoning, long-term, 137
Poisson's equation, 102, 107
Poisson's ratio, 73, 85
polar coordinates, 53
polynomial expansions, high-order, 202
post-doctoral fellows, 1
Powertech Labs, 80
Prandtl number, 29, 162
precipitating agent, 200, 226
preconditioning for conjugate-gradient
    method, 219
predators, 130
pressure wave, 253, 255
pre-stress, 69
primary sequence, 200
primary time scales, 211
print station, double, 273
print, color, 272
printing press, 270, 271
process optimization, 199
product, duality, 246
production of glass, 181
propagating flame front, 254, 256
propagating wave, 285
propane, 253, 255
protein crystal growth, 199
protein macromolecule, 210
protein matrix, 99
pseudophenomenological modeling, 241
pulse, sound, 255
pulses, 97

quadrature, Gaussian, 219
quasi–static motion, 235, 237, 259
quasi-stationary boundary condition, 222
quasi-stationary jet, 256
quasi-steady limit, 32
quasi-steady state, 279

rabbit damage, 116
radiation, 268
radiation equation, 181–183
radiation, spherical, 255
radiative flux, 175, 183, 184, 194
radiative heat transfer problem, 160, 182
radiative intensity, two scale, 191
rate of loading, 229
ratio of specific heats, 27
ray tracing, 181, 187, 188, 194
reaction kinetics, 161
reaction rate, 139, 143, 162
reaction zone, 171, 173, 174
reaction, endothermic, 160

reaction, exothermic, 137, 160
reactions, chemical, 135, 137, 143, 160, 162,
    163
real gradient mapping, 247
reduced Peclet number, 29
reduced Reynolds number, 19, 29
refractive index, 183, 195
refractometry, 203
refrigerant, 22
relaxation distance, 166
relaxation modulus function, 240
reservoir, 199
residual strain, 81
resonance, 67, 70
reversible process, 257
Reynolds number, 50, 255
Reynolds number, particle, 162
Reynolds number, reduced, 29
RHD (Rabbit Haemorrhagic disease), 115, 116
rice grains, 97
Rivlin SEF model, 231
Rockwell International, 270
roller surface finish, 272
roller, static, 272
rolls, tile, 272
Rosseland approximation, 189–191, 194
rotary cooking, 98
rotating chemical bond, 200
rotational inertia, 280
RPI, 135, 271
rubber, 229
rubber bearings, 229
rubbers, highly filled, 249
Runge–Kutta method, 151, 188, 218, 266

saddle point, 149
salt, 203
scattering, isotropic, 197
Schott Glas, 181, 195
scroll compressor, 22
seals, 229
second-order reaction kinetics, 161
second-order scheme, implicit, 219
SEF (Strain Energy Function), 230
semitransparent medium, 181, 182
sensing head, 67
separation of variables, 51
sequence, genetic, 199, 200
seropositive, 130
shape memory alloys, 229
shear viscosity, 26
shear, simple, 230
sheet (of paper), 270
shock waves, 6
SIAM, 1
silica, 229
silicon, 46

simple extension, 230, 233
simple harmonic motion, 72
simple shear, 230
simply supported boundary condition, 75
simulation, numerical, 37, 136, 199
simultaneous partial differential equations, 160
single-pellet dynamics, 161
sink, of acoustic energy, 258
site, active, 200
slenderness parameter, 28
slip angle, 280
slow flow equations, 18, 50
slow time scale, 211
slow viscous flow, 46
smart vibration suppression, 229
smelting furnace, 47
SMM, 1
soaking time, 113
software packages, commercial, 181
solid angle, 183, 187
solid catalyst pellets, 160
solution existence, 245
solution, aqueous, 199
solution, formal, 188
solution, weak, 246
solvability condition, 31
sonic velocity, 259
soot, 184
sound pressure level, 254
sound pulse, 255
sound speed, 256
Southampton University, 271
spalling, 177
spans (of paper), 270
specific gravity, 69
specific heat, 26, 162, 183
spectral band, 186
spectral-element method, 219
spectral radiative intensity, 183, 189
spectrum, electromagnetic, 182
spectrum, opaque, 182
specular boundary condition, 196
speed of light, 183
spherical radiation, 255
SPL, 254
splines, linear, 235, 237, 248
splitter, 273
spontaneous evaporation, 224
spot light count, 118
spring constant, 71
spring–mass–dashpot system, 236
squeeze film, 29
starch crops, 97
starch granules, 99
state space, 236
static friction, 277
static roller, 272

steel lined tank, 90
Stefan condition, 8
Stefan–Boltzmann constant, 163, 190
step function, Heaviside, 177
stiff chemical reaction terms, 135
stiff equation solver, 235
stoichiometric mixture, 253
stoichiometry, 140
strain energy function (SEF), 230
strain history, 229
strain invariants, 231
strain, Green–St. Venant, 232
stress displacement relations, 85
stress field, internal, 182
stress tensor, Piola–Kirchhoff, 233
stress, nominal, 233
stress, true, 233
stress–strain curve, 81
strong coupling, 36
Strouhal number, 50
structural biology, 199
Study Groups with Industry, 1
sum-factorization method, 219
superficial velocity, 166
supersaturation, 201
supersaturation kinetics, 227
surface concentration of reacting species, 137
surface reaction, 176
susceptible, 118
system of algebraic equations, 265
system of hyperbolic equations, 278
Søderburg electrode, 47

tail pipe, 135
tar, 48
Tayler, Alan, 1
Taylor series, 164
team work, 3
technology transfer, 1
temperature, activation, 164
tension (of newspaper web), 271
tension equilibration, 281
tension, thermal, 181
tensioner, 274
tensor-product sum-factorization method, 219
tensor, Cauchy-Green, 232
tensor, Piola–Kirchhoff, 233
tests, cyclical, 236
thermal boundary layer, 162
thermal conductivity, 166, 183, 204
thermal conductivity, effective, 162
thermal contraction, non-homogeneous, 181
thermal enclosure, 203
thermal energy, 27, 33, 256
thermal equilibrium, 162, 176
thermal runaway, 146
thermal tension, 181

thermodynamics, fundamental law of, 257
Thiele modulus, 177
three-dimensional radiative equation, 181, 182
threshold density, 115, 130
tile rolls, 272
time scale, 211, 279
time scales, multiple, 199
Timoshenko theory, 233
tires, 229
toasted grains, 98
trajectory, monotone decreasing, 174
transparent region, 182
transport mechanisms, 201
transport, diffusive, 205
travelling wave, 120
true stress, 233
tubers, 97
turbulent compressible mixing, 254
turbulent jet, 258
turning points, 242
two-dimensional shell theory, 95
two-scale analysis, 190
two-scale radiative intensity, 191

U.S. Government specification, 137
Uncle Tobys Wahgunyah plant, 98
uniform expansions, 170
uniformly valid solution, 216, 218
unique periodic solution, 36
universal gas constant, 27, 164
University of British Columbia, 80
unstable node, 149

vane, 67
vapour diffusion, 199, 200
vapour pressure, 220
vapour pressure, interface, 226
vapourization enthalpy, 207
variational projection operator, 219
vector space, 235
vector, fly, 118
vector, human, 117
vector, insect, 117
vector, wind, 117, 119
velocity potential, 71, 73
velocity test, 49
velocity, interfacial, 225
velocity, sonic, 259
velocity, superficial, 166
vena-contracta effect, 255
vibration suppression, smart, 229
viral disease, 116
virtual mass, 71
virus seeding points, 132
viscometer, 49
viscosity test, 48
viscosity, gas, 162

viscous boundary layer, 162
viscous effects, 280
viscous shear, 66
viscous starch paste, 99
void fraction, 139
volumetric expansion coefficient, 204
vortex shedding, 255, 260, 268
vorticity, 71

Wardang Island, 116
warm-up, 136
washcoat, 135
wave equation, d'Alembert's solution of, 260
wave equation, one-dimensional, 260
wave speed, 124, 128, 130
wave split, 125
wave, infection, 118
wave, outgoing, 261
wave, pressure, 253, 255
wave, propagating, 284
wave, travelling, 120
wave, weakly nonlinear, 267
waves in newspaper web, 274, 275
waves, elastic, 281
weak solution, 246
weak solution, global, 247
weakly nonlinear wave, 267
web, 270
web constitutive equation, 282
web creases, 271
web slip angles, 284
web tension, 271
well-posed PDE, 245
well-posed problem, 58, 245
wetting front, 97
wheat grains, 97
wild rabbits, 116
writing skills, 2

x-ray diffraction, 200

Young's modulus, 73, 85, 277

zeroth-order expansion, 192